ENGINEERS'
GROWTH
REDEFINED

Expertize Uniquely, Communicate Powerfully, and Monetize Effectively in the AI Era

By

PRADEEPKUMAR K PADMANABHAN

DEDICATED TO

Dr. APJ Abdul Kalam

Who Can Benefit The Most From This Book?

Every engineer, regardless of their experience, has the potential to grow beyond their current state. Growth is not just about moving up the career ladder—it's about expanding possibilities, unlocking new opportunities, and creating a fulfilling life.

Through my experience, I have identified seven categories of engineers who can embark on this transformation. Each of them brings unique strengths, motivations, and aspirations to this journey.

Engineers Who Can Start This Journey:

1. Working Engineers

Many engineers with stable jobs feel limited by their work environment. While employment offers financial security, it often comes with constraints on income, growth, and creative freedom.

By redefining their growth, working engineers can:

- Enhance their value in their workplace by demonstrating greater impact.

- Create additional income sources through expertise-driven projects.

- Explore passion-driven work beyond their daily responsibilities.

- Prepare for greater independence instead of relying solely on employment.

Whether they aim to grow within their job or build something beyond it, engineers can amplify their influence and financial success through a structured approach.

2. Engineering Students

Most engineering students spend years studying without clear direction for their future. Instead of waiting until graduation to make decisions, they can start early by:

- Identifying their interests and strengths.

- Gaining hands-on experience by working on real-world problems.

- Building skills beyond academics to stand out in their field.

- Exploring earning opportunities while still studying.

By taking charge of their growth early, students can transition smoothly into a fulfilling career instead of settling for whatever comes their way.

3. Fresh Engineers

After graduation, many engineers struggle to find the right path. Some take any available job without real interest, while others remain uncertain about where to start. A few explore freelancing or contract work, but without guidance, they struggle to grow.

By following a structured approach, fresh engineers can:

- Build credibility and visibility in the industry.

- Attract better opportunities instead of chasing them.

- Gain clarity on their career direction and purpose.

- Develop skills that make them stand out in the job market.

Instead of waiting for the right opportunity, they can create it themselves by taking a proactive approach.

4. Freelance Engineers

Freelancers often have technical expertise but lack structure in their work. Without a clear system, they face challenges such as:

- Inconsistent income due to project-based work.

- Limited scalability beyond individual efforts.

- Difficulty in establishing long-term credibility.

By redefining their growth, freelance engineers can:

- Systematize their expertise into a stable business.

- Create multiple revenue streams beyond one-time projects.

- Expand their reach and influence to attract higher-value opportunities.

A structured approach ensures that freelancers don't just work for income—they build something sustainable.

5. Engineering Entrepreneurs

Entrepreneurs with technical backgrounds often focus on products, services, or manufacturing, but many struggle to expand their reach and maximize their revenue. Their biggest challenges include:

- Lack of effective marketing strategies.

- Dependence on traditional business models.

- Limited visibility in their industry.

By redefining their growth, entrepreneurs can:

- Position themselves as thought leaders in their field.

- Leverage digital platforms to scale their business.

- Monetize their expertise beyond physical products.

Growth for entrepreneurs isn't just about business expansion—it's about leveraging knowledge and innovation for greater impact.

6. Retired Engineers

Many retired engineers have decades of experience, yet they often feel disconnected from the engineering world after retirement. Some seek ways to stay active, while others look for meaningful engagement.

By redefining their growth, retired engineers can:

- Share their expertise with the next generation.

- Explore new opportunities beyond traditional employment.

- Stay relevant and connected with their industry.

Rather than slowing down, they can continue making an impact in new and fulfilling ways.

7. Technical Teachers & Trainers

Educators and trainers play a crucial role in shaping future engineers. However, many rely on traditional methods, which limit their reach and income.

By redefining their approach, they can:

- Expand beyond classroom teaching through digital platforms.

- Reach a larger audience beyond their institution.

- Monetize their expertise effectively through structured systems.

This ensures that their impact goes beyond a single institution and reaches a global audience.

FOREWORD

It is with great honor that I introduce this remarkable book, *Engineers' Growth Redefined*, authored by my esteemed student, Pradeepkumar KP.

Over the years, I have had the privilege of witnessing his unwavering dedication to engineering excellence, personal growth, and his mission to empower engineers to achieve both professional mastery and personal success.

In today's rapidly evolving world, traditional career paths are being reshaped by technological advancements, particularly artificial intelligence (AI) and automation. Engineers are no longer confined to conventional roles; they now have the power to leverage their expertise in ways that were unimaginable just a decade ago. This book serves as a guiding light for engineers navigating this dynamic landscape.

Pradeepkumar presents a structured approach to career growth through three essential pillars—Expertise, Communication, and Monetization. He meticulously breaks down the journey into practical steps:

- Determine, Develop, and Deploy your expertise to build a strong foundation.

- Say, Sell, and Serve by effectively communicating your value to the world.

- Make, Manage, and Multiply wealth by monetizing your expertise strategically.

What sets this book apart is its emphasis on adaptability in the age of AI. Engineers must not only refine their skills but also integrate emerging technologies into their careers. This book equips readers with actionable insights on how AI can enhance their learning, automate processes, and expand opportunities for business and career growth.

Beyond technical skills, this book also reinforces the importance of personal branding, financial literacy, and entrepreneurship—areas that many engineers overlook in their professional journey. The world needs engineers who are not just problem-solvers but also innovators, thought leaders, and contributors to a sustainable future.

Pradeepkumar's approach is both practical and transformational, making this book an essential read for every engineer—whether a student, a working professional, or an aspiring entrepreneur. By applying the principles in this book, engineers can unlock unprecedented career growth and financial independence while making a meaningful impact on society.

I wholeheartedly recommend *Engineers' Growth Redefined* to anyone seeking to navigate the future of engineering with confidence and purpose.

With Love and Best wishes

Dr. Prakash Subramaniam, PhD

Higher Education Leader. Former Dean, Sathyabama Institute of Science & Technology, Chennai

PREFACE

Dear Engineers,

Like many of you, I once believed that securing admission to an engineering college was the gateway to a stable, successful life. I thought that four years of study would guarantee me a well-paying job, and with that job, I could achieve everything I ever wanted. But life had other lessons in store for me.

The reality of my engineering journey was different. I worked in government, public, and private organizations, explored self-employment, and even ventured into entrepreneurship. Yet, despite my technical skills and professional experience, I felt something was missing. Managing money was a constant struggle, and financial stability seemed elusive. More than that, I realized I wasn't truly free—I was living reactively, driven by external circumstances rather than my own terms.

It wasn't until I crossed 50 that I had a profound realization: I had spent decades mastering engineering but had neglected to engineer my own life. This awakening led me on a journey of self-discovery, where I uncovered the principles of personal growth and developed habits that transformed my thinking. I documented these insights in my first book, *Automate Your Growth*, which focused on personal development as the foundation for professional success.

Through my learning, I recognized that career growth is an essential pillar of personal growth—because financial well-being fuels all other areas of life. Inspired by this realization, I began coaching engineers—ranging from students to retired professionals—on how to leverage their expertise for career acceleration, financial independence, and true freedom.

The result of this journey is this book, *Engineers' Growth Redefined*. In this book, I reveal how engineers can:

Expertize uniquely – Identify and position their skills in a way that stands out.

Communicate powerfully – Build influence and convey their value effectively.

Monetize effectively – Create sustainable income and financial freedom in the AI era.

With the rapid advancement of Artificial Intelligence (AI), even a one-person organization can now compete with large corporations. This shift presents both challenges and incredible opportunities for engineers who know how to adapt. To help you navigate this transformation, I have developed the TRUE Growth Model, where I define engineers as Technical Resource Utilization Experts. This model categorizes all engineering and technical resources into seven key areas, helping engineers align their expertise for maximum leverage in the AI-driven future.

The content of this book is structured and explained to help engineers apply these principles for their personal and professional growth. For those seeking additional guidance, my coaching system offers dedicated support for implementation.

I hope this book serves as a catalyst for your growth—helping you redefine success, take control of your career, and build a life of freedom and fulfillment. The strategies shared here are practical, actionable, and designed to empower you in this fast-changing world.

Wishing you grand success in your journey.

For Engineers' growth and happiness!

Pradeepkumar K Padmanabhan

pradeepdesign@gmail.com

www.pradeepkumarkp.com

TABLE OF CONTENTS

Introduction

Hello Engineer, you may be well aware that the world of engineering is changing rapidly. Gone are the days when a degree and a job would guarantee a stable career for life.

Today, technology is evolving faster than ever, and industries are transforming at an incredible pace. If you don't grow, adapt, and establish yourself as an expert in a specific niche, you risk being left behind.

1. Why Engineers Need to Redefine Growth

The world of engineering is undergoing a massive transformation. The traditional model of career growth—get a degree, secure a stable job, work for decades, and retire—no longer guarantees success.

With the rise of AI, automation, and digital innovation, engineers must adapt, evolve, and redefine what growth means in their careers.

If you're still relying on outdated career paths, you may be limiting your potential.

The future belongs to engineers who can master expertise, communication, and monetization to create a career on their own terms.

Let's break down why growth needs to be redefined and how you can take charge of your engineering journey.

The Traditional vs. New Growth Model for Engineers

Old Model of Growth

- Get a degree → Get a job → Work for 30+ years → Retire

- Linear career path with limited flexibility

- Salary increments depend on employer and industry trends

- Expertise is developed slowly over time

- Growth is limited by promotions and job changes

New Model of Growth

- Build expertise → Communicate value → Monetize skills → Continuous learning

- Flexibility to work on your terms (consulting, freelancing, startups)

- Income growth through multiple revenue streams

- Fast-tracked expertise development using AI and automation

- Career growth based on skill mastery, personal brand, and problem-solving

Engineers who redefine growth shift from being job seekers to becoming experts, industry leaders and problem solvers.

3 Reasons Why Engineers Must Redefine Growth

AI & Automation Are Changing the Game

AI is replacing repetitive engineering tasks.

New engineering roles demand creativity, problem-solving, and adaptability.

Engineers must leverage AI instead of competing with it.

If you don't upgrade, you risk becoming obsolete. The solution? Focus on expertise that AI can't replace—critical thinking, design, innovation, and high-value problem-solving.

Traditional Jobs No Longer Guarantee Stability

The gig economy and freelancing are growing faster than ever.

Companies prefer contract-based specialists over full-time generalists.

Engineers must create their own demand by showcasing expertise.

Instead of waiting for opportunities, engineers must actively create them by developing high-demand skills and building their personal brand.

Growth is No Longer Just About Salary—It's About Freedom

The best engineers don't just work for money; they work for impact, freedom, and purpose.

True growth means having control over your career, income, and time.

Engineers who build their brand and expertise can command premium opportunities.

Growth is about creating choices—whether that means working on meaningful projects, building your own business, or achieving financial independence.

The Engineer's Growth Framework: Master, Communicate, Monetize

To redefine growth, engineers must focus on three pillars:

Master Your Expertise – Determine, develop, and deploy specialized skills.

Communicate Your Value – Say, sell, and serve to build authority.

Monetize Your Knowledge – Make, manage, and multiply money.

By following this framework, engineers can stop depending on external factors (like jobs & promotions) and start taking control of their growth.

Take Action: How Can You Redefine Your Growth Today?

Identify one skill area where you want to become an expert.

Start sharing your knowledge online (blog, LinkedIn, YouTube).

Explore new income streams (consulting, freelancing, digital products).

Growth is in your hands. Will you redefine it or stay stuck in the old model?

2. The Shift from Traditional Careers to Expertise-Based Careers

The engineering landscape has shifted. Earlier, engineers were valued for their broad skill sets and ability to work within a structured corporate hierarchy.

Today, specialization is king. Instead of being a generalist in a sea of engineers, you must be an expert in a focused niche, communicate to create, exchange and deliver value and to prosper with the money you earned using your expertise.

For decades, engineers followed a traditional career path:

Get a degree → Land a job → Work for promotions → Retire

This model worked when industries were stable, jobs were secure, and promotions were predictable. But today, AI, automation, and global competition have disrupted this pattern.

Now, the most successful engineers don't just have jobs—they have expertise.

Instead of waiting for promotions, engineers are creating their own demand by mastering niche skills, building authority, and monetizing their knowledge.

Why the Traditional Career Model is Fading

Limited Job Security – AI and automation are replacing repetitive engineering tasks.

Slow Career Growth – Climbing the corporate ladder takes years (or decades).

Income is Controlled by Employers – Salary hikes depend on company policies.

Lack of Flexibility – Engineers are tied to a specific job, location, and industry.

The traditional model is no longer reliable. Engineers must adapt, evolve, and redefine success.

The New Model: Expertise-Based Careers

Instead of chasing jobs, engineers are now:

Becoming Experts – Specializing in high-value, niche skills.

Building Their Brand – Showcasing expertise through content and thought leadership.

Creating Multiple Income Streams – Consulting, digital products, freelancing, and more.

Working on Their Own Terms – Remote work, freelancing, entrepreneurship.

Growth is no longer about job titles; it's about mastering expertise and monetizing it.

Key Differences Between Traditional and Expertise-Based Careers

Traditional Careers	Expertise-Based Careers
Employer controls salary & growth	Engineer controls income & opportunities
Climbing the corporate ladder	Growing as an authority in a niche
Limited to job roles & promotions	Multiple income streams from expertise
Learning is slow & company-driven	Self-driven, fast-paced learning
Dependence on companies for job security	Financial independence through skill monetization

The Power of Expertise in an AI-Driven World

AI is replacing low-value, repetitive tasks. The only way to stay ahead is to focus on what AI cannot replace:

- Creative problem-solving

- Innovative engineering solutions

- Leadership & strategic thinking

- Deep technical expertise in niche areas

By mastering a specific expertise, engineers can command higher pay, better opportunities, and long-term success.

How to Transition to an Expertise-Based Career

- Expertise Uniquely – Identify a high-value skill in demand.Learn deeply, apply knowledge, and build credibility and create solutions, products, and services, pricing and positioning

- Communicate Powerfully – Properly Say (create awareness) and sell (create value) and serve (create impact)

- Monetize Effectively – make, manage and multiply.

The world doesn't reward generalists. It rewards experts.

Now is the time to shift from being just an engineer to becoming an expert engineer. Are you ready?

3.The Role of AI and Emerging Technologies in Shaping Engineering Careers

Engineers, we are standing at the edge of a massive transformation. AI and emerging technologies are reshaping the engineering landscape faster than ever before. The question is no longer if AI will change your career but how you can adapt and thrive in this new era.

Let's break down the impact, opportunities, and necessary skill shifts that engineers must embrace to stay ahead.

How AI is Transforming Engineering Careers

- Automating Repetitive Tasks – AI-powered tools are taking over calculations, simulations, and routine design work, freeing engineers to focus on higher-value problem-solving.

- Enhancing Product Design & Innovation – Generative AI can suggest design improvements, predict failures, and optimize performance before real-world testing.

- Revolutionizing Manufacturing & Construction – AI-driven robotics, 3D printing, and IoT sensors are reducing errors and increasing efficiency in production processes.

- AI in Decision-Making & Problem-Solving – Engineers are now using AI for predictive maintenance, failure analysis, and real-time data-driven decisions in industries like aerospace, automotive, and energy.

- New Career Paths & Emerging Roles – Traditional engineering jobs are evolving into AI-powered roles such as:

  o AI-Driven Design Engineer

  o Data-Driven Systems Engineer

  o Robotics & Automation Specialist

  o AI and Machine Learning Engineer

Engineers Who Adapt Will Be in High Demand

AI will not replace engineers, but engineers who know how to leverage AI will replace those who don't.

Companies will hire engineers who can:

- Use AI-powered software & tools to speed up development.

- Work alongside AI to solve complex engineering challenges.

- Interpret AI-driven insights to make better decisions.

- Automate processes for efficiency and cost reduction.

If you don't evolve, your role might become obsolete. If you learn how to work with AI, you'll be in demand like never before.

Emerging Technologies That Will Shape Engineering Careers

AI is just the beginning! Several emerging technologies are redefining the way engineers work:

- Machine Learning & Big Data – AI-driven analytics help engineers make better predictions, optimize designs, and automate problem-solving.

- Internet of Things (IoT) – Sensors and smart devices are improving real-time monitoring, predictive maintenance, and system automation.

- Blockchain in Engineering – Ensuring secure data sharing, supply chain transparency, and smart contracts in engineering projects.

- 3D Printing & Advanced Manufacturing – Reducing costs and creating highly customized products on demand.

- Virtual Reality (VR) & Augmented Reality (AR) – Enabling immersive prototyping, training, and remote collaboration for engineers.

- Quantum Computing – Unlocking new possibilities in material science, simulations, and cryptography.

How Engineers Can Stay Ahead in an AI-Driven World

1.Master AI-Powered Tools – Learn to use AI-based engineering software like:

- AutoCAD AI (for automated design)

- ANSYS AI Simulation (for predictive engineering)

- AI-driven coding assistants (for software engineers)

2.Specialize in a High-Demand Niche – AI is creating new domains in engineering. Identify a niche that AI is empowering and become an expert in that area.

3.Develop Computational Thinking – Engineers must understand algorithms, automation, and AI logic even if they're not coding AI themselves.

4.Work on Interdisciplinary Skills – The future engineer is not just a technical expert but also understands:

- Business models (how AI impacts industries)

- Communication (explaining AI-driven decisions)

- Ethics & AI governance (understanding responsible AI use)

5.Create Value Beyond AI – AI can analyze and automate, but human creativity, leadership, and innovation will always be irreplaceable. Engineers must:

- Think critically about AI-generated results

- Apply problem-solving in new and creative ways

- Lead projects that integrate AI with engineering principles

Engineers Who Lead AI, Lead the Future

The choice is yours. AI and emerging technologies are not threats—they are opportunities.

The engineers who learn to leverage AI, adapt to new technologies, and focus on expertise rather than routine work will become leaders in their field.

This book will guide you through this transformation step by step, ensuring that you move from being just an engineer to an industry expert and a high-value professional.

Moreover, AI is not here to replace engineers—it's here to amplify our capabilities.

If you learn how to use AI tools effectively, you can increase your efficiency, automate repetitive tasks, and make smarter decisions.

- This book will also show you how to leverage AI in:

- Developing your expertise – AI-driven learning platforms, automated design tools, and smart analytics

- Communicating your value – AI-powered content creation, marketing automation, and customer engagement

- Monetizing your skills – AI-assisted business models, financial management, and scaling strategies

Instead of fearing AI, I want you to embrace it and use it as a tool to accelerate your career growth.

Your Growth Journey Starts Now

This book is your roadmap. It will help you:

- Find and validate your expertise

- Communicate and sell your value to the world

- Monetize your skills and achieve financial freedom

Remember, engineers are problem-solvers by nature. But to thrive in today's world, you must not only solve problems but also position yourself as the expert who can provide the best solutions.

I invite you to take this journey with me. Let's redefine what growth means for engineers like us.

Let's build expertise. Let's create impact. Let's achieve true success.

********** EXPERTIZE **********

CHAPTER 1

UNDERSTANDING EXPERTISE & CAREER GROWTH

1.1. The Power of Expertise in an AI-Driven World

Dear Engineer,

Let's talk about the biggest shift happening in the engineering world today.

AI is no longer the future—it's the present. Machines are designing, analyzing, predicting, and even making decisions. Automation is replacing repetitive tasks, AI-powered simulations are cutting down development time, and intelligent algorithms are handling complex calculations with zero fatigue.

So, the question is: Where do YOU stand in this AI-driven world?

Are you going to be replaced by AI? Or will you leverage AI to amplify your expertise and become indispensable?

The answer depends on whether you are just another engineer or an expert problem solver.

Let me be blunt—generalists will struggle, but experts will thrive.

The more refined and high-impact your expertise is, the more valuable you become. In fact, AI isn't a threat to experts—it's an enabler.

But if you're simply "going with the flow," doing what everyone else is doing, and not defining your niche, you risk becoming irrelevant.

The Power of Expertise: Why Being "Good Enough" is No Longer Enough

In the past, just having an engineering degree and a few years of experience could guarantee a stable career. But today, the industry doesn't reward effort—it rewards impact.

- A mechanical engineer who understands AI-driven automation is more valuable than one who only knows traditional manufacturing.

- A software engineer who specializes in AI security is more in demand than a generic coder.

- A civil engineer who integrates IoT and AI for smart infrastructure is way ahead of someone who just designs buildings.

The trend is clear—depth beats breadth. You must move from being a generalist to a specialist and from a specialist to an expert.

Experts don't compete. They dominate.

The Three Phases of Building Expertise

Expertise isn't something you're born with. It's something you build. And it happens in three phases:

1.Determine Your Expertise – Find your niche. Don't be a jack-of-all-trades; be a master of one. Identify a problem worth solving and go deep.

2.Develop Your Expertise – Learn, apply, refine, and teach. Expertise isn't just about knowledge; it's about execution and results.

3.Deploy Your Expertise – Make your expertise visible. Package it into solutions, communicate it effectively, and let the world know why you are the go-to person in your domain.

AI is Not Your Competition—It's Your Greatest Tool

A lot of engineers fear that AI will take away jobs. That's wrong thinking. AI is here to take away repetitive tasks, not expertise.

Instead of fighting AI, learn to work with it. Use AI to supercharge your expertise.

- Are you an electronics engineer? AI can help you design circuits faster.

- Are you a software developer? AI can optimize your code, suggest solutions, and automate debugging.

- Are you a structural engineer? AI-powered simulations can predict stress points before you even build a prototype.

AI won't replace engineers. But engineers who know how to use AI will replace those who don't.

Your Expertise is Your Career Insurance

If you master a niche, build authority, and apply AI as a tool, your career is future-proof.

This book will guide you through every step of that journey.

- How to find and validate your expertise niche

- How to develop high-value skills that AI cannot replace

- How to package, communicate, and ready to monetize your expertise

The world is shifting fast, and you need to redefine your growth.

1.2 The Three Phases: Determine, Develop, Deploy

Engineers, if you want to thrive in today's AI-driven world, you need to go beyond just having knowledge—you need expertise. Expertise is not just about what you know, but how effectively you apply, refine, and share that knowledge. And to do this, you must go through three essential phases: Determine, Develop, and Deploy.

1. Determine – Find Your Niche, Define Your Path

Before you start mastering a skill, you need clarity on what you should focus on. Many engineers make the mistake of trying to be good at everything. But the real game-changer is specialisation. Specialisation creates authority, and authority creates opportunities.

Here's how you determine your expertise:

1. Self-Analysis: Where are you now? What are your strengths, weaknesses, and interests?

2. Define Your Purpose: What kind of impact do you want to create? Where do you see yourself in five years?

3. Explore Engineering Categories: AI, automation, IoT, mechanical design, software development—pick an area that excites you.

4. Microniche Selection: Narrow it down. Instead of just "AI," choose "AI for Predictive Maintenance in Manufacturing."

5. Validate Your Niche: Test it with the 10-Activity Framework ,5 Why & 5P Method (Passion, Potential, Problem-solving ability, Profitability, and People's demand).

6. Strategy Sheet: Document everything to get crystal-clear direction on your expertise focus.

Once you determine the right path, the next phase begins.

2. Develop – Master the Skill, Build Authenticity

Once you have clarity on your niche, you need to develop expertise in a structured way. Learning alone isn't enough—you need to implement, teach, and refine your skills.

Here's how you develop your expertise:

1. **Learn from the Best:** Follow industry leaders, take structured courses, get mentors.

2. **Apply & Experiment:** Work on real-world projects, create prototypes, test your solutions.

3. **Teach & Share:** Start blogging, making videos, or teaching others—it strengthens your knowledge.

4. **Gather Feedback:** The market will tell you where you need improvement. Take feedback seriously and refine your skills.

5. **Build Your Authenticity:** Your uniqueness lies in your approach. Develop a unique problem-solving mindset.

6. **Leverage AI Tools:** AI can help in rapid learning and automation—use it to gain an edge.

Once you have developed your expertise, it's time to share and monetize it.

3. Deploy – Create, Package, Price, Position, and Ready to Deliver

The real power of expertise comes when you deploy it in a way that creates impact and generates revenue. Having knowledge is one thing; using it to build something valuable is another.

Here's how you deploy your expertise:

1. **Create & Package Products:** Develop eBooks, courses, consulting services, or software solutions.

2. Design & Price Your Offers: Create different pricing strategies based on the value you provide.

3. Leverage AI for Content Creation: Use AI tools to automate course delivery, content generation, and engagement.

4. Use Online Platforms: Sell through an LMS, webinars, video courses, or live workshops.

5. Optimize for Growth: Automate repetitive tasks, scale through collaborations, and keep upgrading your offerings.

6. For physical products, first package and sell the knowledge of their usage and problem-solving benefits before launching the actual product. This will significantly boost sales.

By following these three phases—Determine, Develop, Deploy—you will not just have a job; you will have a career that thrives even in the face of AI disruption.

Your expertise will be your currency in the new digital economy.

So, take a step today. Start by determining your expertise. Then develop it to mastery.

And finally, deploy it to create value and wealth.

Are you ready to redefine your growth as an engineer?

1.3 How AI and Automation Are Reshaping Engineering Roles

Engineers, we are at a pivotal moment in history. The way we work, create, and innovate is being transformed by artificial intelligence (AI) and automation. If you haven't already noticed, the engineering landscape is shifting rapidly, and those who embrace these changes will thrive, while those who resist may find themselves left behind. Let's dive deep into how AI and automation are redefining engineering roles and what you can do to stay ahead of the curve.

1. AI as Your Engineering Assistant, Not Replacement

First, let's bust a myth—AI is not here to replace engineers. Instead, it acts as an amplifier of our abilities, automating repetitive tasks, enhancing decision-making, and providing insights that were previously impossible to obtain. AI-driven tools are now capable of performing complex calculations, running simulations, analyzing big data, and even predicting failures in engineering systems. But remember, AI lacks the creativity, intuition, and ethical judgment that human engineers bring to the table. Your role as an engineer is evolving, not disappearing.

2. Automating Routine Tasks to Free Up Creativity

In the past, engineers spent a lot of time on manual calculations, drafting, and data entry. Now, automation tools handle these routine tasks, allowing engineers to focus on high-value problem-solving. For example:

- Civil Engineers use AI-powered software for structural analysis, optimizing designs in real-time.

- Mechanical Engineers leverage automation in CAD software to speed up prototyping.

- Software Engineers utilize AI-driven code generators to write boilerplate code, freeing up time for strategic thinking.

- Electrical Engineers apply AI in circuit design, reducing human error and enhancing efficiency.

With automation taking care of the grunt work, your role becomes more about strategic thinking, innovation, and interdisciplinary collaboration.

3. AI-Powered Decision Making

AI is revolutionizing how decisions are made in engineering projects. With the help of machine learning algorithms, engineers can analyze large datasets, detect patterns, and make data-driven decisions faster than ever before. Some examples include:

- Predictive Maintenance: AI analyzes sensor data to predict when machinery or infrastructure will fail, reducing downtime and maintenance costs.

- Smart Manufacturing: AI-driven robotics optimize production lines, minimizing waste and maximizing efficiency.

- Energy Optimization: AI helps engineers design more energy-efficient systems by analyzing power consumption patterns.

- Autonomous Systems: Self-driving cars, drones, and robotic process automation (RPA) are reshaping industries, with engineers leading the way in their development and deployment.

4. Engineering Specializations Are Evolving

With AI and automation taking over certain tasks, some traditional engineering roles are transforming, while new specializations are emerging. If you're an engineer looking to future-proof your career, consider diving into these evolving fields:

- AI & Machine Learning Engineering: Engineers are developing AI models for automation and decision-making.

- Data Science & Analytics: Engineers who can interpret and act on data insights will be in high demand.

- Robotics & Automation Engineering: Designing, building, and maintaining automated systems will become a core function.

- Cybersecurity for Engineers: As automation grows, securing systems from cyber threats is more critical than ever.

- Sustainable & Green Engineering: AI is helping optimize renewable energy and environmental impact.

If you're still working in a traditional engineering role, now is the time to explore how AI can enhance your expertise. The key is to *adapt* rather than resist.

5. The Engineer's Role in AI Development

One crucial fact to remember is that AI systems don't build themselves—engineers do. From developing AI algorithms to integrating them into real-world applications, engineers play a leading role in AI innovation. Whether you're designing AI-powered manufacturing processes, creating smart infrastructure, or coding AI models, your expertise is essential in making AI work effectively.

6. Upskilling for the AI Era

Now that we've established how AI and automation are reshaping engineering roles, what should you do next?

- Embrace Continuous Learning: Enroll in AI, machine learning, and automation courses. Platforms like Coursera, Udemy, and specialized engineering academies offer targeted learning programs.

- Develop an AI Mindset: Start thinking about how AI can enhance your engineering projects. Be proactive in implementing AI tools.

- Master Data Handling: Engineers who can work with data—collecting, analyzing, and interpreting it—will have a competitive edge.

- Experiment with AI Tools: Get hands-on experience with AI-driven software relevant to your industry, whether it's MATLAB, AutoML, TensorFlow, or AI-powcred CAD tools.

- Stay Updated: Follow industry trends, attend webinars, and engage in AI-focused engineering communities.

Engineers as the Architects of the AI Revolution

The future of engineering is exciting, and AI is at its core. Instead of fearing automation, leverage it to maximize your impact. Your job as an engineer is not just to adapt but to *lead* in this era of AI-driven transformation. Whether you're optimizing existing systems, designing intelligent automation, or pioneering new AI applications, the opportunities are limitless.

Are you ready to redefine your role as an engineer in the AI age? The choice is yours. Embrace the change, master AI, and position yourself as a leader in the new world of engineering!

Chapter 2

Determining Your Expertise

2.1 Self-Analysis: Where Are You Now?

Before you can grow, before you can master your expertise, and before you can build a high-income engineering career—you need clarity. You need to pause and ask yourself, "Where am I right now?"

Think of this as the foundation of your career transformation. Just like an engineer doesn't start designing a structure without analyzing the soil and terrain, you shouldn't start building your expertise without understanding your current position.

So, let's break it down.

1. Your Current Skillset & Experience

Take an honest look at your technical and non-technical skills. Make a list of:

- The technologies, tools, and software you are comfortable with.

- The projects you've worked on and the problems you've solved.

- Any certifications, courses, or specialized training you have.

Now, ask yourself:

- Are my current skills relevant to the industry trends?

- Do I have expertise that is future-proof in the age of AI and automation?

- Which areas do I enjoy working in the most?

If your skills are outdated or scattered across different fields, don't worry.

The first step toward expertise is recognizing where you stand.

2. Your Interests & Passion in Engineering

Passion matters. You might have the skills, but do you enjoy what you do?

The key to sustainable growth is aligning your expertise with what excites you.

Ask yourself:

- Which engineering problems do I love solving?

- What kind of work excites me the most?

- If money wasn't a factor, what would I love to work on?

The intersection of your skills and passion will help you identify your ideal domain.

3. Your Purpose & Long-Term Vision

Every successful engineer has a WHY behind their work. Why are you in this field?

What do you want to achieve in the next 5, 10, or 20 years?

To define your purpose, answer these:

- What kind of impact do I want to create as an engineer?

- What kind of career growth do I want—corporate leadership, entrepreneurship, consulting, or freelancing?

- Do I want to work with cutting-edge technologies, innovate, or teach others?

Your long-term vision will guide your journey toward expertise.

4. Your Strengths & Weaknesses

No one is perfect. You need to know your strengths and weaknesses to create a roadmap for growth.

- Strengths – What are you naturally good at? What do people appreciate about your skills?

- Weaknesses – What areas need improvement? Are there knowledge gaps that could hold you back?

Being self-aware allows you to leverage your strengths and work on your weaknesses effectively.

5. Your Market Value & Opportunities

Even if you love a particular skill, you must ensure it has market demand. Research and answer:

- What are the high-demand engineering skills today?

- Is there a growing need for my expertise in the industry?

- Who are the top engineers in my niche, and what are they doing differently?

This will help you determine whether your expertise needs refinement or a shift toward an in-demand specialization.

The Four Key Areas of Self-Analysis

To determine where you are now, assess yourself in these four areas:

A. Technical Expertise – What do you already know?

- What are your core technical skills?

- Which engineering problems can you confidently solve?

- How updated are you on the latest trends in your industry?

- What certifications, projects, or experiences validate your expertise?

Example:

- A mechanical engineer might have strong CAD skills but lacks expertise in AI-driven simulations.

- A software engineer might be skilled in Python but needs to learn machine learning frameworks.

Your goal: Identify where your expertise is strong and where you need improvement.

B. Career Position & Industry Demand – Are you aligned with market needs?

- Are your skills in demand in your industry?

- Is your current job role future-proof, or is it becoming obsolete?

- How has AI and automation impacted your engineering field?

- Are you in a job that excites you and aligns with your career goals?

Example:

- A civil engineer in traditional construction might notice AI-driven smart cities are the future and needs to upgrade skills.

- An electrical engineer working in manual circuit design sees that AI-powered PCB automation is growing fast.

Your goal: Ensure your current skills align with the future needs of your industry.

C. Personal & Professional Growth – Are you progressing?

- Are you growing consistently in your field?

- Do you feel stuck, stagnant, or unsatisfied in your career?

- Are you continuously learning and upgrading your skills?

- Do you have a mentor or network that helps you grow?

Example:

- If you've been in the same job role for years without significant learning, it's time to upskill.

- If you struggle to explain or sell your expertise, improving communication skills is necessary.

Your goal: Identify areas where growth is slow and create a strategy for consistent learning and improvement.

D. Financial Stability & Monetization Potential – Are you making the most of your skills?

- Are you financially stable with your current income?

- Do you have multiple sources of income from your expertise?

- Have you considered monetizing your engineering skills (consulting, teaching, product development)?

- Are you investing in yourself (books, courses, training)?

Example:

- A software engineer with a full-time job but no passive income might explore AI-based freelancing or digital product creation.

- A mechanical engineer could create an online course on design simulations to earn additional revenue.

Your goal: Identify how to increase your income by leveraging your expertise.

The Self-Assessment Framework: 5-Level Rating

To make self-analysis easier, use this rating system for each of the four areas:

Area	Level 1 (Beginner)	Level 2 (Basic)	Level 3 (Intermediate)	Level 4 (Advanced)	Level 5 (Expert)
Technical Expertise	Limited skills	Basic knowledge	Can handle real-world projects	Strong expertise	Industry leader
Career Growth	No growth	Slow progress	Steady growth	Fast-tracking career	Top of industry

Personal Development	No learning	Occasional learning	Regular upskilling	Structured learning	Always evolving
Financial Growth	Struggling	Just managing	Stable income	Multiple income streams	Financially free

Action Plan: If you rate yourself below Level 3 in any area, focus on improving that area through learning, projects, and networking.

The Clarity Sheet – Your Self-Analysis in Action

Create a simple self-analysis clarity sheet and answer these:

What are my core engineering skills?

What are the gaps in my expertise?

How relevant are my skills to the current industry trends?

Am I satisfied with my career growth?

Am I continuously learning and improving?

How stable is my financial growth?

Am I monetizing my expertise effectively?

Your goal: Answering these questions honestly will give you a clear roadmap for growth.

What's Next?

Self-analysis is the first step toward mastering your expertise. Once you know where you are now, the next step is defining your purpose and long-term goals to shape your engineering career.

Once you have clarity on where you stand today, it's time to take action. Start documenting your findings. Create a Personal Expertise Blueprint with:

- Your current skills and strengths

- Your passion and interests

- Your long-term vision

- The gaps you need to fill

- Rate yourself using the self-assessment framework.

This self-analysis will help you Determine, Develop, and Deploy your expertise in the best possible way.

Now, let's move to the next step—Defining Your Purpose and Long-Term Goals.

Remember, clarity is power. Once you know where you stand, you can take control of where you're going.

2.2 Defining Purpose and Long-Term Goals

Why Do You Need a Purpose?

As an engineer, you're trained to solve problems. But here's a question—what problem are you solving in your own career?

Many engineers start their journey following a standard path:

- Get a degree.

- Find a job.

- Gain experience.

- Earn a good salary.

But at some point, a realization hits: "Is this all there is?"

That's why defining your purpose is crucial. Your purpose gives meaning to your work.

It keeps you motivated when challenges arise. It helps you build a career, not just do a job.

If you don't define your purpose, you might end up working hard but feeling directionless. So, let's take control and design a career with intent.

Step 1: Discover Your Purpose

Your purpose is the driving force behind your career decisions. It's not just about making money—it's about creating impact.

Ask yourself these powerful questions:

1. What excites me the most in engineering?

2. What kind of problems do I enjoy solving?

3. If I had unlimited time and money, what would I work on?

4. Who do I want to help with my expertise?

5. How do I want to be remembered as an engineer?

Example:

- If you love automation, your purpose could be: *To simplify industrial processes and improve efficiency through automation.*

- If you enjoy teaching, your purpose could be: *To mentor young engineers and bridge the gap between academia and industry.*

- If you're passionate about sustainability, your purpose could be: *To create eco-friendly engineering solutions for a greener planet.*

Your purpose is the foundation of your career. Once you define it, every decision will align with it.

Step 2: Define Your Long-Term Goals

Having a purpose is great, but without clear goals, it's just a dream. Let's break it down into long-term career goals.

1. Where Do You Want to Be in 10 Years?

Visualize your future. Answer these:

- What kind of projects do I want to work on?

- Do I see myself in a leadership position, running a business, or becoming a top expert in my field?

- What impact do I want to create in the engineering world?

This big vision will be your North Star.

2. What Are Your 5-Year Milestones?

To reach your 10-year goal, what should you achieve in 5 years?

- Master a specific technology?

- Become a recognized expert in your niche?

- Build a profitable side business?

- Transition into a high-paying role?

Example: If your 10-year goal is to be a top AI-driven manufacturing expert, your 5-year goal could be:

- Get certified in AI & automation.

- Work on real-world automation projects.

- Start sharing knowledge through blogs, courses, or mentorship.

3. What Are Your 1-Year Action Plans?

To reach your 5-year milestone, what must you achieve this year?

- Learn a high-value skill?

- Start networking with industry leaders?

- Take on a leadership role?

- Build an audience or start a side hustle?

Example: If your 5-year goal is to become a recognized engineering consultant, your 1-year goal could be:

- Build expertise in your niche.

- Start sharing your insights on LinkedIn or YouTube.

- Connect with professionals who need consulting services.

Step 3: Align Your Goals with Your Purpose

Now, check if your goals align with your purpose. If they don't, adjust them. Your goals should:

- Reflect your strengths.

- Be connected to the problems you love solving.

- Move you closer to your dream career.

Write Down Your Purpose & Goals

Document your answers. This is your Personal Career Blueprint. Keep revisiting it and refining it over time.

Purpose Gives You Power

When you have a clear purpose and long-term goals, you stop feeling stuck. You wake up with excitement because you know exactly what you're working towards.

2.3. The Role of AI in Skill Development & Personalized Learning

Engineering is Evolving—Are You?

As engineers, we thrive on learning and adapting to new challenges. But let's be honest—keeping up with the latest skills, technologies, and trends can feel overwhelming.

This is where AI changes the game.

AI is no longer just a buzzword—it's a powerful tool that can accelerate skill development and personalize learning like never before. Instead of spending years figuring things out, AI can customize your learning path, suggest relevant resources, and even provide real-time feedback.

If you're not using AI to upskill, you're leaving massive opportunities on the table.

1. AI as Your Personal Learning Assistant

Imagine having a personal mentor available 24/7—guiding you, recommending resources, and testing your knowledge. That's exactly what AI-powered learning platforms are doing today.

- Adaptive Learning Platforms like Coursera, Udemy, and LinkedIn Learning use AI to analyze your strengths and weaknesses and suggest personalized courses.

- AI Chatbots & Tutors (like ChatGPT and Google Gemini) can answer your questions instantly, explain complex concepts, and even generate code examples.

- Customized Learning Paths: AI analyzes your past learning habits and creates a unique syllabus tailored to your needs.

Example:

If you're an automotive engineer wanting to shift into AI-driven vehicle automation, AI can recommend courses on machine learning, computer vision, and IoT in automotive—based on your current skill level and industry trends.

2. AI-Driven Skill Validation & Practical Learning

Learning theory is not enough—you need hands-on experience. AI helps engineers validate their skills through:

AI-Powered Simulations & Virtual Labs

- Tools like MATLAB, Simulink, and Ansys now use AI to simulate real-world scenarios. You can test your designs and ideas without expensive physical prototypes.

Real-Time Coding & Debugging Assistants

- AI-powered tools like GitHub, Copilot and Tabnine predict your code, suggest fixes, and help you learn new programming languages on the go.

AI in Engineering Design & Prototyping

- Platforms like Autodesk Fusion 360 and SolidWorks now use AI to automate part design, optimize material usage, and reduce errors.

Example:

- If you're a civil engineer working on smart cities, AI-powered simulation tools can help test structural integrity, optimize energy consumption, and predict material wear—before construction even begins.

3. AI-Powered Personalized Learning for Engineers

AI personalizes your learning based on your goals, current expertise, and industry demand. Here's how:

AI-Based Career Guidance

- AI tools like IBM SkillsBuild and LinkedIn Career Explorer analyze market trends and recommend skills that are in high demand for your industry.

- If you're an electrical engineer, AI might suggest learning edge AI chips and embedded systems for the future of smart devices.

AI-Generated Study Plans & Schedules

- Platforms like Notion AI, ChatGPT, and Jasper can create structured learning plans based on how much time you can dedicate.

- If you want to master industrial automation in 6 months, AI can break it into weekly modules with specific resources.

AI-Powered Language & Communication Training

- If you're aiming for global career opportunities, AI-driven tools like Duolingo and Grammarly help polish your communication skills in different languages.

Example:

An engineer in India planning to work in Germany can use AI-based language training to become fluent in technical German within months—without traditional classes.

4. AI-Driven Networking & Mentorship

Engineers grow faster when they connect with the right mentors and communities. AI can help with that too!

AI-Based Mentor Matching

- Platforms like ADPList and LinkedIn use AI to connect you with industry experts based on your career goals.

AI-Powered Knowledge Sharing

- AI curates relevant forums, research papers, and discussion threads, so you stay updated without spending hours searching.

Automated Webinars & Workshops

- AI-generated transcripts, summaries, and real-time translations make global learning accessible.

Example:

If you're a mechanical engineer specializing in robotics, AI can help connect you with industry leaders, suggest robotics hackathons, and even provide summaries of cutting-edge research in your field.

5. The Future: AI as an Engineering Partner

As AI continues to evolve, it will do more than just assist in learning—it will become an engineering partner that:

- Identifies knowledge gaps and suggests real-time solutions.

- Automates repetitive tasks, so engineers focus on innovation.

- Enables faster prototyping with AI-assisted simulations.

- Enhances collaboration with AI-driven project management tools.

AI isn't here to replace engineers—it's here to supercharge your growth. The sooner you leverage AI for skill development, the faster you'll become an industry leader in your field.

Stay Ahead or Fall Behind

The engineering industry is evolving at the speed of AI. The question is—are you evolving with it?

Your next step: Start using AI-powered learning tools today!

- Identify a high-value skill to learn.

- Use AI to create a personalized study plan.

- Apply your knowledge with AI-driven simulations and projects.

Remember: The future belongs to engineers who master AI—not those who fear it.

2.4. Selecting and Validating a Microniche (10 Activity Framework & 5P Method & 5 WHY)

Why Do Engineers Need a Microniche?

As an engineer, you have skills and knowledge in a broad field. But if you try to serve everyone, you end up serving no one effectively. The real power lies in specialization—focusing on a microniche where you can be the go-to expert.

A microniche helps you:

- Stand out in a crowded industry

- Become an authority in a specific area

- Attract high-value clients or job opportunities

- Command higher earnings

For example, instead of being a general mechanical engineer, you can niche down into AI-driven robotics for industrial automation.

Now, how do you select and validate your microniche? We'll use:

10 Activity Framework – To explore and shortlist potential microniches.

5P Method – To validate if your chosen microniche is viable.

5 WHY – Ask why this niche?

Step 1: Selecting Your Microniche – 10 Activity Framework

This framework helps you discover your true expertise and interest in a structured way.

1.Self-Reflection on Strengths & Interests

- What are you naturally good at?

- What technical problems excite you the most?

- Which topics do you enjoy discussing or working on?

Example: If you love renewable energy and AI, you might explore AI-driven energy optimization systems.

2.Assess Market Demand

- Are industries investing in this field?

- Is there a growing need for engineers in this niche?

- Are there job postings, businesses, or startups focusing on this area?

Example: Edge AI for IoT devices is in high demand with the rise of smart industries.

3.Identify Industry Trends & Future Scope

- Is the technology evolving, or is it becoming obsolete?

- What innovations are happening in this space?

- Will this niche still be relevant in 5-10 years?

Example: Traditional manufacturing is declining, but smart factories powered by AI and IoT are booming.

4.Analyze Competition & Existing Solutions

- Who are the top players in this field?

- Are there gaps in the solutions being offered?

- Can you offer something unique or better?

Example: If you notice that most AI-powered industrial automation tools are expensive, you could create affordable AI-based predictive maintenance solutions.

5.Leverage Your Past Experience

- What past projects or work experience can help in this niche?

- Do you already have skills that are transferable to this microniche?

Example: If you've worked in embedded systems, you can transition into AI-integrated embedded devices.

6.Test Your Knowledge & Passion

- Can you explain key concepts in this niche to others?

- Are you excited to research and work on it daily?

Example: If explaining machine learning in edge computing excites you, it might be a great microniche.

7.Define Your Target Audience

- Who will benefit from your expertise?

- Are they companies, startups, students, or professionals?

Example: If you specialize in energy-efficient AI processors, your target audience could be IoT manufacturers and AI startups.

8.Identify Monetization Opportunities

- Can you create products, courses, or consulting services around this niche?

- Are there companies willing to pay for expertise in this field?

Example: If you're skilled in autonomous drones, you can offer drone automation consulting or training programs.

9.Connect with Experts & Communities

- Are there professional groups, LinkedIn communities, or research forums in this niche?

- Are experts talking about this field?

Example: Edge AI for medical diagnostics has active communities discussing trends and research.

10.Prototype & Validate with a Small Project

- Can you build a small project or prototype in this niche?

- Can you write a blog, make a video, or share insights to gauge interest?

Example: If you create a YouTube video on AI in IoT devices and it gains traction, it confirms market interest.

By going through these 10 activities, you'll shortlist a few strong microniche options. Now, let's validate them using the 5P Method.

Step 2: Validating Your Microniche – 5P Method

Once you have a potential microniche, run it through the 5P method to ensure it's a profitable and sustainable choice.

1.Passion – Do you love working on this niche?

Are you genuinely interested in this topic?

Will you enjoy learning and working in this niche long-term?

Example: If you love AI for medical imaging, you won't get bored working on AI-driven diagnostic solutions.

2.Problem-Solving – Does it solve real-world problems?

Are businesses, industries, or individuals struggling with challenges in this field?

Can you provide valuable solutions with your expertise?

Example: Many companies struggle with AI implementation in manufacturing. If you can simplify AI integration, it's a powerful niche.

3.Profitability – Can you make money in this niche?

Are companies, clients, or students willing to pay for expertise in this field?

Are there consulting, product, or service opportunities?

Example: AI-powered supply chain optimization can save businesses millions, so companies are willing to invest in experts.

4.People & Audience – Is there a target market?

Are there specific industries, professionals, or companies looking for solutions in this niche?

Can you reach and connect with them easily?

Example: Automated drone surveillance for agriculture has a well-defined market (agribusinesses, government agencies).

5.Practicality – Can you execute and grow in this niche?

Do you have the necessary skills, resources, or experience?

Is it realistic to enter this niche and grow?

Example: If quantum computing excites you but requires years of specialized study, it may not be immediately practical. But AI-driven cybersecurity could be a practical alternative.

Step3: Validating Your Microniche – The 5 Why Method

Finding a niche is important, but validating it ensures long-term success. Many engineers struggle because they pick a field based on assumptions rather than deep clarity. The 5 Why Method helps you dig deeper into your microniche to ensure it aligns with your expertise, passion, and market demand.

How Does the 5 Why Method Work?

It's simple—ask "Why?" five times to uncover the true reason behind your chosen microniche. Let's break it down with an example:

Step 1: Ask "Why you should select this niche?" (First Layer)

"I want to specialize in AI-driven automation for manufacturing."

Because I want to focus on AI-driven automation,

Step 2: Ask "Why?"

"Why do I want to focus on AI-driven automation?"

Because industries are moving towards smart factories, and I see opportunities here.

Step 3: Ask "Why?" Again

"Why do I see opportunities in smart factories?"

Because traditional manufacturing processes are inefficient and need optimization.

Step 4: Keep Digging Deeper

"Why do manufacturers need better optimization?"

Because high operational costs and human errors reduce productivity and profitability.

Step 5: Get to the Core

"Why is solving this problem important for me?"

Because I have expertise in industrial automation, and I enjoy designing intelligent systems that enhance efficiency.

Final Validation:

This process clarifies that your microniche isn't just about trends—it aligns with your expertise, passion, and real-world needs.

Why This Works for Engineers?

Removes Guesswork – Ensures you're not choosing a niche just because it's popular.

Aligns with Strengths – Validates that your skills and interests support long-term growth.

Confirms Market Demand – Helps you assess if industries are willing to pay for your expertise.

Strategy Sheet for Clarity

To solidify your choice, fill out a simple strategy sheet with:

- Your Chosen Microniche: (Example: AI-driven predictive maintenance for industrial automation)

- Problem You're Solving: (Factories face unexpected machine failures, leading to downtime)

- Target Audience: (Manufacturers, industrial engineers, automation companies)

- Your Unique Value Proposition: (An AI model that predicts failures & reduces downtime by 40%)

- Monetization Plan: (Consulting, online training, SaaS product, etc.)

This sheet will give you clarity on whether to move forward.

Your Microniche = Your Authority

By selecting and validating a strong microniche, you position yourself as an expert instead of just another engineer.

- Follow the 10 Activity Framework to explore your strengths, market demand, and future trends.

- Use the 5P Method to validate if your niche is sustainable, profitable, and exciting.

- Strengthen the logic of its importance by applying the 5 WHYs technique.

- Fill out the Strategy Sheet to confirm your focus area.

Once you have clarity, start building your expertise, creating content, and offering solutions in your microniche!

So, what's your microniche? Let's refine it together!

2.5.Strategy Sheet for Clarity

Once you've identified your potential microniche, the Strategy Sheet helps you gain absolute clarity about your direction. This ensures that you focus on an area that aligns with your passion, problem-solving ability, market demand, and monetization potential.

Here's a structured format you can use to finalize your microniche strategy:

Step 1: Define Your Microniche

What is your chosen niche?

Example: AI-driven Predictive Maintenance for Industrial Automation

Why did you select this niche?

Example: Industries lose millions due to unexpected equipment failures. AI can predict breakdowns, reducing downtime.

What excites you about this field?

Example: Combining AI with mechanical engineering to create real-world impact.

Step 2: Identify the Problem You're Solving

What are the top 3 pain points in this niche?

- Manufacturers face unplanned machine failures leading to production loss.

- Traditional maintenance is costly and inefficient.

- Engineers struggle to implement AI-driven solutions.

How does your expertise help solve these problems?

Example: I can create AI models that analyze machine data, predict failures, and optimize maintenance schedules.

Step 3: Define Your Target Audience

Who will benefit from your expertise?

- Manufacturing companies

- Industrial automation engineers

- Factory owners

- Maintenance teams

Who is your ideal client or employer?

Example: Mid-to-large-scale manufacturers investing in smart factories.

What are their biggest challenges?

Example: High maintenance costs, lack of AI expertise, inefficient operations.

Step 4: Your Unique Value Proposition (UVP)

What makes you different from others in this niche?

Example: Most AI maintenance solutions are complex. I simplify AI integration for manufacturers.

What is your unique offering?

- AI-powered maintenance strategy consulting

- Custom-built predictive maintenance AI models

- Training programs for engineers to implement AI-driven solutions

Step 5: Monetization Plan

How will you make money from your expertise?

- Consulting Services – Helping industries implement AI-driven maintenance

- Online Courses – Teaching engineers how to use AI for predictive maintenance

- Product Development – Creating AI-powered maintenance software

- Affiliate Marketing – Partnering with companies selling maintenance tools

What are your pricing strategies?

- Entry-Level: Free content (blogs, videos, webinars)

- Mid-Level: Paid workshops, consulting calls

- High-Level: Customized AI solutions, full-scale training programs

Step 6: Future Scalability

How can you scale your expertise in the next 3-5 years?

Example:

- Build a personal brand as an AI-maintenance expert

- Develop a software-as-a-service (SaaS) platform

- Expand globally by targeting international industries

What are the next 3 actions you will take?

- Create content (articles, LinkedIn posts) on AI in predictive maintenance.

- Engage with industry professionals and forums.

- Launch a free webinar to attract potential clients.

Example for Final Clarity Statement

"I specialize in AI-driven Predictive Maintenance for Industrial Automation. I help manufacturers reduce downtime and costs by implementing AI-based solutions through consulting, training, and digital products. My goal is to become a recognized authority in this field and scale my impact through technology and education."

Your Strategy Sheet = Your Roadmap

- This strategy sheet is your roadmap to becoming an expert in your microniche.

- It ensures you stay focused and aligned with your goals.

- Once filled, you have a clear, actionable plan for developing, communicating, and monetizing your expertise.

Ready to finalize your microniche? Fill out your strategy sheet today and take the first step toward your career transformation!

CHAPTER 3

DEVELOPING YOUR EXPERTISE

3.1. Learning Through Mentors & Structured Education

As an engineer, continuous learning is non-negotiable. Technology evolves rapidly, and staying relevant requires structured education and mentorship. The right mentor and learning system can shortcut your growth, helping you avoid unnecessary trial and error.

Why Learning Through Mentors is Essential?

Imagine two engineers starting their careers at the same time:

Engineer A learns everything on their own, struggling to find the right resources.

Engineer B follows a structured path guided by an experienced mentor.

Who do you think will progress faster?

Engineer B will reach expertise much sooner because of direct guidance and clarity.

A mentor is someone who has already walked the path you want to take. They help you:

- Avoid common mistakes

- Gain practical, industry-specific knowledge

- Get insider insights that no textbook will teach

- Expand your network and connect with the right people

- Stay accountable and motivated

How to Find the Right Mentor?

- A great mentor should be:

- An expert in your niche with real-world experience

- Willing to share knowledge openly

- Able to challenge and push you beyond your comfort zone

- Accessible through online courses, books, webinars, or personal coaching

Where to Find Mentors?

- Follow industry leaders on LinkedIn & Twitter

- Join professional groups and online communities

- Enroll in mentorship programs & masterclasses

- Read books and listen to podcasts from experts

- Attend conferences and networking events

Pro Tip: Don't just "find" a mentor—earn their mentorship by showing dedication and curiosity. Engage with their content, ask thoughtful questions, and apply their lessons.

Structured Education: The Fast-Track to Expertise

Mentorship alone is not enough—you need structured education to complement hands-on learning.

3 Key Ways to Learn Efficiently:

1. Online Courses & Certifications

- Platforms like Coursera, Udemy, MIT OpenCourseWare, and edX offer world-class courses.

- AI-driven platforms personalize learning based on your progress.

- Certification boosts your credibility in the job market.

Pro Tip: Don't just take courses—apply what you learn immediately through projects or by teaching others.

2. Books, Research Papers & Blogs

- Books provide deep, structured knowledge in a single resource.

- Research papers keep you updated on cutting-edge advancements.

- Blogs and newsletters from industry leaders share real-time insights.

- Must-Read Books for Engineers:

- *The Lean Startup* – For innovation & entrepreneurship

- *The Fourth Industrial Revolution* – To understand future technologies

- *AI Superpowers* – How AI is transforming industries

Pro Tip: Create a reading habit—20 minutes a day can transform your knowledge over time.

3.Hands-on Learning (Projects & Internships)

- Theoretical knowledge means nothing without practical application.

- Work on real-world projects to develop skills faster.

- Contribute to open-source projects to gain visibility and experience.

- Internships and freelancing offer on-the-job learning with mentorship.

Pro Tip: Keep a portfolio showcasing your projects, achievements, and contributions. This attracts better opportunities.

Action Plan: Your Learning Roadmap

1.Identify your niche and find 2-3 mentors in the field.

2.Enroll in relevant online courses and work towards certifications.

3.Dedicate 20-30 minutes daily to reading books, blogs, and research papers.

4.Work on real-world projects and document your learnings.

5.Engage with mentors and communities—ask questions and seek feedback.

Remember: "Learning never stops for an engineer." The more you learn, apply, and share, the faster you grow.

3.2. Applying Knowledge Through Real-World Projects

Learning theory is essential, but real expertise comes from application.

Engineers who actively apply their knowledge through real-world projects develop deeper skills, build confidence, and gain practical experience that sets them apart.

Why Real-World Projects Matter?

Imagine an engineer who has spent years studying but has never built, tested, or deployed a real project. Compare that to an engineer who learns by doing—building prototypes, solving real problems, and optimizing solutions.

Who do you think is more valuable in the industry?

The one who has applied knowledge and demonstrated their skills in real-world scenarios.

Benefits of Working on Real-World Projects:

- Bridges the gap between theory and practice

- Helps you develop problem-solving skill

- Builds confidence in your abilities

- Creates a portfolio to showcase your expertise

- Increases chances of job offers and freelancing opportunities

- Helps you identify your strengths and improve weaknesses

Pro Tip: Every concept you learn should be applied in a project—this is the fastest way to master skills.

How to Apply Your Knowledge?

1.Solve Real Problems in Your Industry

- Identify challenges in your industry and create a technical solution.

- Research existing solutions and improve them with your own innovation.

- If you work in a company, automate a task or optimize a process.

Example:

- A mechanical engineer could design an energy-efficient system for a factory.

- A software engineer could develop an AI-powered chatbot for customer service.

Action Step: Identify one problem in your industry and brainstorm a project idea.

2.Contribute to Open-Source Projects

- Open-source communities allow you to collaborate with experts.

- You gain hands-on experience working on real-world applications.

- Your contributions improve your credibility and visibility.

Where to Start?

- Explore GitHub or GitLab for projects in your field.L

- Look for beginner-friendly issues and start contributing.

- Join developer forums like Stack Overflow, Dev.to, or Medium.

Action Step: Find an open-source project that matches your niche and contribute.

3.Work on Freelance or Internship Projects

- Practical experience from internships or freelance gigs enhances your portfolio.

- Real clients give you real-world constraints and requirements.

- You develop business and communication skills along with technical expertise.

Platforms to Find Projects:

- Upwork, Fiverr, Freelancer (for freelance work)

- LinkedIn Jobs, Internshala (for internships)

- Kaggle (for AI and data science challenges)

Action Step: Create a profile on a freelance or internship platform and apply for 3 projects.

4.Build Personal & Passion Projects

- Your own project allows you to experiment and innovate.

- It gives you complete creative control over the solution.

- Passion projects help you stand out in job interviews and networking.

Examples of Passion Projects:

- A home automation system using IoT (for electronics engineers).

- A self-learning chatbot using AI (for software engineers).

- A low-cost water filtration system (for civil engineers).

Action Step: Start a personal project that excites you and challenges your skills.

5.Join Hackathons & Competitions

- Hackathons provide real-world challenges under time constraints.

- You collaborate with experts and learn teamwork and problem-solving.

- Winning hackathons boosts your credibility and opens new career opportunities.

Where to Find Hackathons?

- Devpost (software and AI challenges)

- Kaggle (data science competitions)

- NASA Space Apps Challenge (engineering & innovation)

- MIT Solve Challenges (impact-driven tech solutions)

Action Step: Search hackthon for engineers in India.Register for a hackathon or technical competition in your field.

Action Plan: Apply Your Knowledge Now!

- Choose a real-world problem in your field.

- Pick a project type: Open-source, freelance, personal, or hackathon.

- Start small: Work on a minimum viable product (MVP) and improve it over time.

- Document your project: Write a blog, create a GitHub repository, or share on LinkedIn.

- Get feedback & iterate: Improve based on user input.

Remember: "Knowledge without application is wasted." The more projects you work on, the faster you grow.

3.3. Teaching and Sharing Expertise to Reinforce Learning

"The best way to learn is to teach." – Have you ever explained a complex topic to someone and realized that you understood it even better afterward? That's the power of teaching and sharing expertise.

For engineers, mastering a skill isn't just about acquiring knowledge—it's about applying, refining, and teaching it to others. When you share what you know, you:

- Strengthen your own understanding

- Build credibility in your niche

- Create a network of like-minded professionals

- Open doors to new career and business opportunities

So, let's explore how you can start teaching and sharing your expertise effectively.

Why Teaching Reinforces Learning?

Cognitive Boost: Teaching forces you to organize your thoughts, making your understanding sharper.

Feedback Loop: When you explain concepts, you identify gaps in your own knowledge.

Deep Mastery: To teach, you must simplify and clarify complex topics, which leads to better retention.

Creative Thinking: Different learners ask different questions, pushing you to explore new angles.

Pro Tip: If you can't explain a concept simply, you haven't mastered it yet.

How to Teach and Share Expertise Effectively?

1.Start by Writing & Documenting What You Learn

Writing forces clarity. Whether through blog posts, LinkedIn articles, or documentation, putting your thoughts into words solidifies your learning.

Where to Share?

- LinkedIn Posts – Short insights & lessons

- Medium/Dev.to Blogs – In-depth articles

- GitHub Repositories – Document technical projects

- Personal Website – Build your professional brand

Action Step: Pick a recent learning and write a short LinkedIn post explaining it in simple terms.

2.Create Simple Video Tutorials & Demos

Many engineers prefer visual learning. You don't need fancy equipment—just start with your phone or screen recording software.

Content Ideas:

- Explaining a Concept – E.g., "What is AI Bias and How to Fix It?"

- Live Coding or Design Demo – Walk through a project step by step.

- Common Mistakes & How to Avoid Them – Help beginners by sharing insights.

Where to Share?

- YouTube – For structured tutorials

- Instagram Reels & TikTok – For short, engaging lessons

- LinkedIn & Twitter – For quick insights and micro-content

Action Step: Record a 2-minute video explaining a simple concept and share it online.

3.Host Webinars & Live Sessions

Live interaction allows you to answer questions in real time and build stronger connections.

Where to Host?

- Zoom, Google Meet, Microsoft Teams – For structured live classes

- LinkedIn Live & YouTube Live – For public sessions

- Clubhouse & Twitter Spaces – For audio-based discussions

Action Step: Plan a 15-minute live session to teach a small topic in your niche.

4.Mentor & Guide Others

Nothing accelerates growth like mentorship. By guiding others, you develop leadership skills while reinforcing your expertise.

Ways to Mentor:

- One-on-One Mentorship – Help a junior engineer grow.

- Join a Mentorship Program – Platforms like ADPList & MentorCruise.

- Internship/Apprenticeship Programs – Take interns under your guidance.

Action Step: Find a junior engineer or student and offer to mentor them on a small project.

5.Build a Community Around Your Expertise

A community creates a support system where learning and teaching happen naturally.

How to Build One?

- Create a WhatsApp or Telegram Group – Share daily insights & resources.

- Start a LinkedIn Group – Bring engineers together.

- Launch a Discord/Slack Channel – Host discussions & Q&A sessions.

Action Step: Start a Telegram or WhatsApp group with 5-10 engineers in your niche.

Teaching = Growth + Impact + Monetization

When you consistently share your expertise, you don't just learn better—you also:

- Build authority in your field

- Expand your professional network

- Attract career opportunities

- Create income sources (courses, coaching, consulting)

"Your knowledge becomes valuable when you share it." So, start teaching today and watch your expertise skyrocket!

3.4. Building Authenticity and Credibility

As an engineer, your expertise alone won't set you apart—authenticity and credibility will. In an age where AI and automation are reshaping industries, trust and genuine expertise are more valuable than ever.

So, what do authenticity and credibility mean for an engineer?

- Authenticity – Being real, transparent, and true to your values.

- Credibility – Proving your expertise through experience, results, and trustworthiness.

Why does it matter?

- People trust real experts, not just those who claim to be one.

- Credibility attracts opportunities, whether jobs, clients, or collaborations.

- Authenticity creates long-term influence, not just short-term attention.

1. Be True to Your Expertise & Experience

Your credibility starts with what you know. Don't fake expertise—build it.

How to do it?

- Stick to what you truly know – Don't try to be an expert in everything.

- Acknowledge limitations – It's okay to say, "I don't know, but I'll find out."

- Keep learning – Continuous learning enhances your credibility.

Action Step: Write down three areas where you already have strong expertise and one area where you need improvement.

2. Showcase Real Work & Results

Talk is cheap—proof matters.

Whether you are an engineer, consultant, or entrepreneur, credibility comes from what you've done.

Ways to showcase expertise:

- Project Portfolios – Share your best work (GitHub, Behance, LinkedIn, personal website).

- Case Studies – Explain how you solved real-world engineering problems.

- Certifications & Degrees – Highlight relevant qualifications but combine them with practical proof.

Action Step: Share a real case study on LinkedIn about a problem you solved.

3. Teach & Share Valuable Insights

One of the fastest ways to build credibility is to educate others. When you teach, you position yourself as an authority.

Where to share?

- LinkedIn, Medium, Dev.to – Write about engineering solutions.

- YouTube & Podcasts – Create technical explainers.

- Webinars & Workshops – Host live training sessions.

Action Step: Create a short LinkedIn post explaining a concept in your field.

4. Be Consistent in Your Messaging

People trust those who show up consistently. Whether it's content, communication, or leadership—be present.

How to maintain consistency?

- Post valuable insights regularly – Even once a week builds credibility.

- Engage in discussions – Answer questions and contribute to conversations.

- Deliver what you promise – Whether deadlines or commitments, stick to them.

Action Step: Commit to posting one valuable insight per week on your platform of choice.

5. Build Social Proof & Testimonials

- People trust experts recommended by others. Social proof builds instant credibility.

Ways to get social proof:

- Client Testimonials – Ask happy clients or colleagues to share their experiences.

- Recommendations on LinkedIn – Request peers, mentors, and clients to endorse your skills.

- Speaking Engagements & Guest Features – Be interviewed on podcasts, write guest blogs.

Action Step: Ask one colleague or client to write a LinkedIn recommendation for you.

6. Be Transparent & Ethical

Trust is hard to earn but easy to lose. Engineers must uphold ethics, transparency, and honesty in their work.

How to stay authentic?

- Give credit where it's due – Acknowledge mentors and sources.

- Don't overpromise – Set realistic expectations.

- Be honest about mistakes – Transparency builds long-term trust.

Action Step: Reflect on a mistake you made and share a lesson learned from it.

Authenticity + Proof = Influence

If you want to stand out, don't just claim expertise—prove it with real actions.

- Stay true to your values

- Showcase real work

- Educate and share knowledge

- Be consistent and ethical

"Your expertise makes you knowledgeable. Your authenticity makes you relatable. Your credibility makes you trusted."

3.5. The Role of AI in Skill Development & Personalized Learning

Engineers, we are living in an AI-driven era where learning is no longer restricted to textbooks, universities, or traditional training programs.

Artificial Intelligence (AI) is changing how we acquire, refine, and apply skills, making learning more personalized, efficient, and results-driven.

But here's the real question: Are you leveraging AI to accelerate your learning and expertise? Let's dive into how AI is reshaping skill development and what you can do to stay ahead.

1. AI-Powered Personalized Learning Paths

Gone are the days of one-size-fits-all learning.

AI adapts to your needs and tailors a customized learning experience based on your current skill level, interests, and career goals.

How AI personalizes learning:

- Adaptive Learning Platforms – AI analyzes your strengths and weaknesses, recommending specific topics and resources.

- Smart Course Recommendations – Platforms like Coursera, Udemy, and LinkedIn Learning use AI to suggest courses based on your learning patterns.

- AI-Driven Skill Assessments – Tools evaluate your skills in real time and offer insights on where to improve.

Action Step: Try AI-powered learning platforms like Skillsoft Percipio, LinkedIn Learning, or Coursera to get tailored recommendations.

2. AI-Driven Virtual Mentors & Chatbots

Imagine having a 24/7 AI mentor who can guide you, answer your queries, and provide learning resources instantly.

How AI acts as a mentor:

- Chatbots & AI Tutors – Tools like ChatGPT, Socratic by Google, and IBM Watson answer technical queries in real time.

- AI-Powered Code Assistants – GitHub Copilot, CodeWhisperer, and Tabnine help engineers learn to code more effectively.

- AI-Based Career Guidance – Platforms like Pymetrics analyze personality traits and suggest career paths.

Action Step: Use AI tools like ChatGPT or GitHub Copilot to get instant guidance while learning new technologies.

3. Faster Skill Acquisition with AI-Powered Content

AI is speeding up how we consume knowledge through automated content generation, summarization, and explanation.

AI tools that make learning faster:

- Summarization Tools – Tools like ScribeSense or TLDRthis condense long articles into key insights.

- AI-Powered Video Learning – YouTube's AI-generated transcripts help engineers quickly find relevant content.

- Speech-to-Text & AI Notes – Otter.ai and Notion AI convert lectures and meetings into structured notes.

Action Step: Use AI-powered summarization tools to extract key insights from long research papers or technical blogs.

4. Hands-On Learning with AI Simulations & VR

AI is not just helping us read and watch—it is enabling real-time, hands-on learning through simulations, virtual labs, and augmented reality (AR).

AI-powered practical learning tools:

- Simulated Environments – Engineers can practice skills in digital twins and AI-generated scenarios.

- AI-Driven Coding Sandboxes – Platforms like Replit and LeetCode use AI to provide instant feedback on coding exercises.

- Virtual Reality (VR) Training – Engineers can now learn complex machinery operations using AI-enhanced VR systems.

Action Step: Explore AI-powered coding assistants or VR-based engineering simulations to enhance practical learning.

5. AI for Continuous Learning & Skill Tracking

Learning doesn't stop after one course or certification. AI is enabling continuous skill tracking and upskilling based on industry trends.

How AI helps with continuous learning:

- Real-Time Industry Insights – AI tools track evolving trends and suggest skills that will be in demand.

- AI-Driven Skill Gaps Analysis – AI platforms evaluate engineers' skills and recommend growth areas.

- Automated Learning Reminders – AI-based apps keep track of progress and suggest regular upskilling opportunities.

Action Step: Use AI-driven career platforms like LinkedIn Learning or IBM SkillsBuild to monitor your progress and stay updated.

AI is Your Learning Accelerator

AI is not here to replace engineers—it's here to help engineers learn smarter, faster, and more effectively.

The real winners will be those who embrace AI to upskill, specialize, and stay ahead of the industry curve.

- Use AI to personalize your learning journey

- Leverage AI-powered mentors and skill trackers

- Embrace hands-on AI simulations and virtual training

- Stay updated with AI-driven industry insights

Remember: "The engineers who master AI-powered learning will be the ones shaping the future of technology."

3.6. Gathering Feedback and Refining Your Expertise

As an engineer, your expertise is not a static achievement—it's a continuous journey of improvement.

One of the most powerful accelerators in this journey is feedback. The right feedback helps you refine your knowledge, improve your skills, and stay relevant in an ever-evolving industry.

But let's be real—not all feedback is useful. Some feedback can be vague, overly critical, or even misleading.

The key is to gather meaningful feedback, analyze it effectively, and use it to sharpen your expertise.

1. Why Feedback is Essential for Growth

Think of feedback as a mirror that reflects your strengths and blind spots. Without it, you might be working hard but not necessarily improving in the right direction.

Key benefits of feedback in expertise-building:

- Helps you identify gaps in your knowledge

- Allows you to improve your problem-solving approach

- Enhances your ability to communicate technical ideas effectively

- Keeps you aligned with industry demands and best practices

- Builds trust and credibility with peers, mentors, and clients

Action Step: Start viewing feedback as a growth tool rather than criticism.

2. Where to Gather Valuable Feedback?

Not all feedback is created equal. You need to tap into the right sources to get actionable insights.

Where to collect high-quality feedback?

- Mentors & Experts – They provide strategic guidance based on experience.

- Peers & Colleagues – They offer practical insights from a shared work environment.

- Clients & Users – Their feedback helps you refine real-world applicability.

- Online Communities & Forums – Platforms like Stack Overflow, GitHub, and LinkedIn give diverse perspectives.

- AI-Powered Feedback Systems – Tools like Grammarly (for writing), Codacy (for coding), and AI tutors provide instant, data-driven insights.

Action Step: Identify 2-3 key sources of feedback based on your field and expertise goals.

3. The Right Way to Ask for Feedback

Getting useful feedback isn't just about asking "What do you think?"—it's about asking the right questions to get clear, actionable responses.

Best ways to ask for feedback:

- Be specific: Instead of "Is this good?" ask "How can I improve this design for better efficiency?"

- Ask for both strengths and weaknesses: This gives a balanced perspective.

- Use structured feedback forms: If possible, use rating scales and comment sections to get detailed insights.

- Follow up with clarifying questions: If feedback is vague, ask "Can you give me an example?"

Action Step: Frame your next feedback request using clear, focused questions.

4. Filtering & Analyzing Feedback Effectively

Not all feedback is worth acting on. Some might be biased, outdated, or irrelevant. The trick is knowing what to keep and what to ignore.

How to evaluate feedback properly:

- Look for patterns – If multiple people mention the same issue, it's worth fixing.

- Consider the source – Feedback from an industry expert carries more weight than random opinions.

- Check for objectivity – Constructive feedback provides suggestions for improvement, not just criticism.

- Compare with industry standards – If feedback misaligns with industry best practices, validate it before acting.

Action Step: Create a simple "Feedback Review Sheet" where you categorize feedback into:

- High-priority improvements

- Minor adjustments

- Irrelevant or non-actionable feedback

5. Implementing Feedback for Continuous Improvement

Feedback is useless if you don't apply it strategically. The goal is to refine your expertise step by step.

How to apply feedback effectively:

- Make a feedback-to-action plan – Prioritize changes based on impact.

- Use iteration cycles – Improve in small steps rather than making drastic changes.

- Test & validate improvements – Check if changes lead to measurable progress.

- Seek follow-up feedback – Continuous refinement is key to long-term expertise.

Action Step: After receiving feedback, set clear action items and track progress over time.

6. Using AI to Automate & Enhance Feedback Loops

AI is transforming how engineers receive and process feedback. Instead of waiting for human reviews, AI-driven tools provide instant, data-backed insights to speed up expertise development.

AI tools for automated feedback:

- For coding: GitHub Copilot, Codacy, DeepSource

- For writing & documentation: Grammarly, Hemingway App

- For presentations & communication: Yoodli (AI speech coach)

- For business & product validation: AI-powered customer analytics tools

Action Step: Integrate at least one AI feedback tool into your workflow for faster learning cycles.

Feedback is the Fastest Way to Expertise

Refining your expertise is not about working harder—it's about working smarter with the right feedback loops. Engineers who embrace feedback as a growth catalyst will evolve faster, build credibility, and create impactful solutions.

- Seek feedback from mentors, peers, and AI-driven tools

- Filter & analyze feedback strategically

- Apply feedback in small, consistent iterations

- Use AI-powered tools to speed up learning & improvement

Remember: "The best engineers are not those who know everything but those who refine their expertise through continuous feedback and adaptation."

CHAPTER 4

DEPLOYING YOUR EXPERTISE

4.1. Product Creation & Content Design

Engineers often excel at solving problems, but when it comes to packaging their expertise into a sellable product, many struggle. The key to sustainable career growth isn't just about knowing your field—it's about turning that knowledge into a structured, valuable product that others can benefit from.

Let's break down the step-by-step process of creating an expertise-driven product and designing compelling content around it.

1. Why Engineers Must Create Their Own Products

Relying solely on a job or freelance work limits your income and impact. When you create a product based on your expertise, you:

- Leverage your knowledge to help a wider audience

- Build credibility as an industry leader

- Create multiple income streams beyond your job

- Gain financial freedom by scaling your expertise

Your product could be:

- An online course teaching specialized engineering skills

- A technical eBook or guide

- A software tool or automation script

- A consulting package offering expert advice

- A video training series

- A physical product or kit that engineers can use

The goal is to convert your knowledge into something valuable that people are willing to pay for.

2. Identifying the Right Product Idea

Your first step is to identify what problem you want to solve. The best products come from deeply understanding a pain point in your industry.

Use this simple framework:

- Problem: What is a common issue engineers in your niche face?

- Solution: How can your knowledge provide an effective solution?

- Audience: Who will benefit the most from your product?

- Value Proposition: Why should they buy from you instead of someone else?

Example:

- Problem: Many mechanical engineers struggle with mastering simulation software.

- Solution: Create an online course on "Mastering Finite Element Analysis (FEA) for Beginners."

- Audience: Mechanical engineers, students, and professionals who want to upskill.

- Value Proposition: Hands-on, project-based learning with industry-relevant case studies.

Action Step: Write down 3-5 product ideas based on your niche expertise and target audience.

3. Designing High-Value Content for Your Product

Once you know what you're creating, the next step is content design—structuring your knowledge in a way that makes learning easy and engaging.

Key Principles of Effective Content Design

- Clarity & Simplicity: Break down complex topics into easy-to-follow lessons.

- Actionable Learning: Include exercises, templates, and real-world examples.

- Engaging Format: Mix text, visuals, videos, and interactive elements.

- AI Integration: Use AI tools to enhance learning (e.g., Chatbots, AI-driven quizzes).

Structuring Your Content

1. Introduction: Explain the problem, the solution, and what users will gain.

2. Fundamentals: Cover essential concepts in a clear and structured manner.

3. Hands-on Application: Provide real-world case studies and exercises.

4. Advanced Techniques: Offer deeper insights and optimization strategies.

5. Conclusion & Next Steps: Guide learners on how to continue improving.

Example for an Online Course:

Module 1: Introduction to Finite Element Analysis (FEA)

Module 2: Understanding Material Properties & Meshing Techniques

Module 3: Running Simulations in ANSYS

Module 4: Analyzing & Interpreting Results

Module 5: Case Study: Structural Analysis of a Mechanical Component

Action Step: Outline your product content into key sections or modules.

4. Choosing the Right Product Format

Your expertise can be delivered in multiple ways. Choose a format that suits your skills and audience preferences.

Digital Products:

- Online Courses (Teachable, Udemy, or AI-powered LMS)

- Ebooks & Guides (PDFs, Notion templates)

- Webinars & Workshops (Live Zoom sessions)

Physical Products:

- Engineering Kits (Tools, components, DIY kits)

- Books & Workbooks (Printed manuals)

Software & Automation:

- AI-driven Solutions (Custom scripts, automation tools)

- SaaS Platforms (Subscription-based software)

Action Step: Select the best format that aligns with your expertise and audience needs.

5. Packaging & Pricing Your Product

Once your product is ready, presentation and pricing matter!

How to Package Your Product for Maximum Appeal

- Professional Branding: Use AI-powered design tools like Canva for stunning visuals.

- Engaging Product Descriptions: Clearly explain who it's for and why it's valuable.

- Bonuses & Extras: Add checklists, templates, and cheat sheets to increase perceived value.

Pricing Strategies for Engineers

Pricing is not just about cost—it's about value. Here's how you can price strategically:

- Entry-Level Pricing: ₹499-₹2,000 for small digital guides or beginner courses.

- Mid-Tier Pricing: ₹5,000-₹15,000 for detailed courses or consulting packages.

- Premium Pricing: ₹25,000+ for high-ticket coaching or custom solutions.

Pro Tip: Test different price points and offer limited-time discounts to drive initial sales.

Action Step: Decide on your pricing strategy and create a compelling product offer.

6. Delivering & Scaling Your Product

Creating a product is just the start—efficient delivery ensures customer satisfaction.

Ways to Deliver Your Product Efficiently

- Use LMS Platforms – Teachable, Kajabi, or AI-powered systems.

- Automate Access – Use Zapier or payment gateways for instant access.

- Provide Excellent Support – Use AI chatbots or email sequences for post-sales support.

Scaling Strategies to Grow Your Product's Reach

- Affiliate Marketing – Partner with influencers in your niche.

- AI-Powered Ads – Run Facebook & Google Ads with AI targeting.

- Community Building – Create LinkedIn groups or private Discord communities.

Action Step: Choose at least one method to deliver and scale your product efficiently.

Build, Launch, and Monetize

Creating your own product is the fastest way to build authority, impact lives, and generate wealth as an engineer.

- Identify a problem and craft a valuable solution

- Design high-quality content that is easy to consume

- Choose the best format for your audience

- Package and price strategically

- Use automation and AI tools to streamline delivery

Remember: Engineers who create their own products don't just survive in the industry—they lead it.

4.2. Packaging, Pricing & Strategy

Once you've created a product based on your engineering expertise, the next crucial step is packaging, pricing, and strategizing for maximum impact. Many engineers make the mistake of focusing only on technical quality while neglecting how the product is presented and monetized.

Let's dive into how you can package your expertise attractively, price it profitably, and position it strategically in the market.

1. Packaging – How to Make Your Product Attractive

Your product might be highly valuable, but if it doesn't look appealing, people won't buy it. Packaging is about how you present, structure, and deliver your product to create maximum perceived value.

Key Elements of Product Packaging

Professional Branding:

- Use clear, eye-catching visuals (Use Canva, Figma, or AI-based design tools).

- Have a consistent color scheme and logo for all materials.

- Keep the design simple, clean, and modern.

Compelling Product Description:

- Clearly define the problem your product solves.

- Highlight key benefits instead of just features.

- Use persuasive language that connects with your audience.

Structured Content:

- Divide the product into clear modules or sections.

- Use a logical flow that makes learning or implementation easy.

- Offer quick wins—valuable takeaways from the first lesson/module.

Bonus Add-ons to Increase Perceived Value:

- Cheat sheets, templates, checklists

- Extra case studies or live Q&A sessions

- Private community access

Example:

If you're selling an AI-driven Mechanical Design Course, instead of just offering videos, package it as:

"AI-Powered CAD Mastery: Learn, Design, and Automate – Complete Bundle"

- 10+ hours of HD training videos

- Downloadable CAD templates

- Free access to AI design automation tools

- Live mentorship calls & Q&A

Action Step: List down 3 key ways you can enhance the packaging of your product.

2. Pricing – How to Charge the Right Amount

Many engineers underprice their expertise due to self-doubt. The reality is that your pricing should reflect the value you provide, not just the effort you put in.

3 Proven Pricing Models

1. Value-Based Pricing (Recommended)

   - Charge based on the transformation you provide, not just time spent.

   - Example: If your course helps an engineer get a ₹10 lakh salary hike, you can charge ₹20,000-₹50,000 easily.

2. Tiered Pricing (Multiple options for different budgets)

   - Basic: ₹4,999 (Core content only)

   - Pro: ₹9,999 (Extra case studies & templates)

   - Premium: ₹24,999 (1-on-1 coaching & mentorship)

3. Competitive Pricing (Pricing relative to the market)

- Research similar courses/products and position yourself at a premium if you provide extra value.

- Example: If others are charging ₹5,000, and you offer a better version with AI tools, you can charge ₹15,000+.

Key Pricing Factors to Consider

- Demand & Urgency: If your expertise is rare or in high demand, charge a premium.

- Your Brand Positioning: Are you seen as an industry expert? Build authority first.

- Customer Willingness: Higher pricing often attracts serious buyers who value quality.

- Scalability: If your product can be sold to many people, pricing can be lower.

Example:

- An AI-based predictive maintenance toolkit could be priced at ₹30,000 if it helps companies save ₹10 lakh annually.

- A Resume & LinkedIn optimization service for engineers could be ₹9,999 if it helps them secure high-paying jobs.

Action Step: Decide on your product's pricing strategy and justify the price based on the value provided.

3. Strategy – How to Position & Sell Your Product Effectively

A great product with bad marketing won't sell. You need a clear sales strategy to position your expertise and convince buyers that your solution is worth their investment.

The 3-Step Sales Strategy for Engineers

1. Build Awareness (Say)

- Leverage LinkedIn, Twitter, and YouTube to create educational content.

- Share real-world case studies and success stories.

- Use AI to automate email marketing and content scheduling.

2. Create Demand (Sell)

- Run webinars/workshops to showcase your expertise.

- Offer free valuable resources (lead magnets) to build trust.

- Use AI-driven chatbots to handle inquiries efficiently.

3. Support & Retain Customers (Serve)

- Provide exceptional post-sale support.

- Offer private community access for buyers.

- Use AI-based CRM tools to follow up and offer personalized upsells.

Package, Price & Sell Like an Expert !

To successfully monetize your expertise, focus on these 3 key areas:

Packaging: Make your product visually appealing, well-structured, and high-value.

Pricing: Charge based on value, not effort—use tiered or premium pricing models.

Strategy: Position your product effectively, create demand, and convert customers.

4.3. Leveraging AI for Content Creation & Product Delivery

AI has revolutionized the way we create, distribute, and deliver content. As an engineer looking to monetize your expertise, integrating AI-driven tools can automate tasks, enhance productivity, and improve user experience.

Let's break down how you can leverage AI for content creation and streamline product delivery to maximize your impact and income.

1. AI for Content Creation – Work Smarter, Not Harder

Creating high-quality content consistently can be time-consuming, but AI can accelerate the process. From writing, video editing, graphic design, and coding—AI tools can do a lot of the heavy lifting.

How AI Helps in Content Creation

AI for Writing & Blogging

- Use ChatGPT, Jasper, or Copy.ai to generate blogs, eBooks, course content, and FAQs.

- AI-powered tools like Grammarly & QuillBot help refine and optimize your writing.

- SEO tools like SurferSEO can boost rankings and drive organic traffic.

AI for Video Content

- Synthesia & Pictory: Convert text into AI-generated videos with voiceovers.

- Descript: Edit videos by editing text (removes filler words & adds captions).

- Runway ML: AI-powered video enhancement, background removal, and effects.

AI for Graphic Design & Presentations

- Canva & Adobe Firefly: Create social media graphics, thumbnails, and ads.

- Beautiful.ai & Tome: AI-powered slide decks and presentations for courses & webinars.

- DALL·E & MidJourney: AI-generated images for marketing materials.

AI for Coding & Automation

- GitHub Copilot & ChatGPT Code Interpreter: AI-powered coding assistance.

- Bubble & Adalo: No-code platforms for building apps with AI automation.

- Zapier & Make.com: Automate workflows like email follow-ups & content scheduling.

Example:

Imagine you're an AI-Driven Mechanical Engineer creating a course on "Generative Design in CAD".

- Use ChatGPT to draft course scripts.

- Use Synthesia to generate AI-powered video lectures.

- Use Canva to create presentation slides & marketing materials.

- Use Notion AI to organize content and brainstorm ideas.

Action Step: Identify 2-3 AI tools that can help you create content faster and more efficiently.

2. AI for Product Delivery – Automate & Scale Your Impact

Once your content is ready, you need an efficient delivery system. AI tools can help you automate sales, onboarding, support, and engagement.

AI-Powered Product Delivery Methods

1. Online Courses & Learning Management Systems (LMS)

- Teachable, Thinkific, and Kajabi: AI-powered course platforms with built-in analytics.

- LearnDash & Podia: LMS tools for structured learning experiences.

- AI-generated quizzes & feedback loops to improve student engagement.

2.AI for Automated Customer Support & Engagement

- Chatbots (ChatGPT, Drift, Intercom): 24/7 AI-powered support.

- AI-powered CRM (HubSpot, Zoho, Freshsales): Personalize responses & automate follow-ups.

- AI email marketing (ConvertKit, MailChimp AI): Segment and nurture leads automatically.

3.AI for Live Webinars & Workshops

- WebinarJam & Demio: AI-assisted webinar hosting with automated replays.

- Otter.ai & Fireflies.ai: AI-driven transcription and note-taking.

- AI coaching bots: Provide automated training suggestions based on user behavior.

4.AI for E-Commerce & Digital Product Delivery

- Gumroad & Payhip: AI-powered platforms for selling digital products.

- Shopify AI: Smart product recommendations & automated upselling.

- Stripe & Razorpay: AI-driven payment processing for seamless transactions.

Example:

If you are selling an AI-powered Resume Optimization Course for Engineers:

- Use Thinkific to host the course with AI-driven assessments.

- Use Intercom AI to handle support and FAQs.

- Use ConvertKit AI to send automated emails and course reminders.

Action Step: Identify 1 AI-powered course platform or delivery tool to automate your content distribution.

3. AI for Personalization – Deliver a Tailored Experience

AI doesn't just automate—it can personalize learning and product experiences based on customer behavior.

AI-Powered Personalization Strategies

1.Adaptive Learning & AI Tutors

- AI tools analyze user progress and suggest next steps dynamically.

- Example: Coursera & Udemy AI suggest personalized learning paths.

2.AI-Powered Recommendation Engines

- Platforms like Amazon & Netflix use AI to suggest content—engineers can too!

- Example: "Based on your interest in Structural Design, check out this FEA Simulation Masterclass!"

3.AI Chatbots for Personalized Assistance

- AI-driven assistants can guide users through course materials or product usage.

Example:

If you're selling a 3D Printing Masterclass for Mechanical Engineers, AI can:

- Track student progress & suggest additional modules.

- Provide personalized AI-based quiz feedback.

- Offer AI-driven coaching bots to answer FAQs.

Action Step: Explore AI tools that personalize learning experiences for your users.

Use AI to Create, Deliver & Scale

As an engineer, leveraging AI in content creation and product delivery will help you:

- Create high-quality content faster and easier.

- Deliver products seamlessly through AI-powered automation.

- Personalize the experience to boost engagement and retention.

Next Steps:

- Identify 3 AI tools that will automate your content creation and delivery.

- Start implementing AI in one area today (content, delivery, or support).

4.4. Online Course Platforms, LMS, and AI-Based Automation Tools

If you are an engineer looking to share your expertise, build a brand, and monetize your knowledge, then setting up an online course is one of the best ways to do it.

Learning Management Systems (LMS) and AI-based automation tools help you streamline content delivery, engage learners, and maximize efficiency.

This section will guide you through the best platforms, AI tools, and strategies to launch and scale your online course.

1. Why Online Courses? The Future of Knowledge Sharing

The online learning industry is booming, with AI-driven platforms making it easier than ever to create and sell courses. Engineers who specialize in niche technical areas can leverage this trend to:

- Generate passive income by selling pre-recorded courses.

- Build credibility and authority in their domain.

- Reach a global audience and impact thousands of learners.

- Automate teaching and customer support with AI tools.

2. Choosing the Right Online Course Platform (LMS)

Hosted LMS Platforms (All-in-One Solutions)

These platforms handle course hosting, sales, marketing, and student engagement—perfect for engineers who want a turnkey solution.

Best LMS Platforms for Engineers

- Teachable – User-friendly, great for beginners, built-in sales & marketing tools.

- Thinkific – Drag-and-drop course builder, customizable sales pages.

- Kajabi – All-in-one platform with advanced automation, community-building features.

- Podia – Affordable option with email marketing & digital product sales.

Example:

If you're a Structural Engineer creating a course on "Finite Element Analysis for Beginners," you can use Thinkific to:

- Host course videos & quizzes.

- Automate sales & student onboarding.

- Provide downloadable resources (PDFs, models, templates).

Action Step: Choose one LMS platform and explore its features.

Self-Hosted LMS (For Full Control)

If you prefer more control over customization, branding, and pricing, then self-hosted LMS platforms are a great option.

Best Self-Hosted LMS Tools

- LearnDash (WordPress Plugin) – Full control over course structure, pricing, and branding.

- Moodle – Open-source, great for universities & large-scale education.

- LifterLMS – Flexible WordPress-based LMS with membership features.

Example:

A Software Engineer creating a Python AI Course can use LearnDash to:

- Host lessons on a WordPress website.

- Offer one-time payments or subscriptions.

- Integrate with WooCommerce & payment gateways.

Action Step: If you want full control, explore LearnDash or Moodle.

3. AI-Based Automation Tools for Online Courses

AI-powered tools can save time, improve student engagement, and personalize learning.

AI Tools for Course Creation

- ChatGPT & Jasper – AI-powered content writing for course materials.

- Synthesia & Pictory – AI-generated video lessons & voiceovers.

- Descript – Edit videos by editing text, auto-remove filler words.

AI Tools for Student Engagement

- Fireflies.ai & Otter.ai – AI-powered transcription for lectures.

- Tidio & Drift AI Chatbots – Automate student inquiries & support.

- Quizlet AI & Kahoot! – AI-generated quizzes & gamified learning.

AI Tools for Sales & Marketing

- ConvertKit AI & MailChimp – Automated email campaigns.

- HubSpot AI CRM – Personalized student interactions.

- SurferSEO & Copy.ai – AI-driven SEO and content marketing.

Example:

A Civil Engineer teaching "AI in Smart Cities" can use:

- Synthesia to create AI-generated video lectures.

- Quizlet AI to build interactive quizzes.

- HubSpot AI to automate email sequences & follow-ups.

Action Step: Identify one AI tool that can automate your course delivery.

4. Monetizing Your Course – Pricing & Delivery Models

Pricing Models

- One-Time Payment – Best for standalone courses (e.g., ₹4999 for a CAD Design Course).

- Subscription Model – Monthly/Yearly payments for access to new lessons & updates.

- Membership Community – Students pay to access content + community support.

- Hybrid Model – Combine self-paced courses with live coaching calls.

Example:

An AI & ML Engineer could offer:

- Basic course at ₹2999 (Self-paced).

- Premium version at ₹9999 (Includes live Q&A sessions).

Action Step: Define a pricing model for your course.

5. Scaling with AI-Based Marketing & Sales Funnels

Once your course is live, you need AI-driven marketing to get more students.

Best AI-Powered Marketing Strategies

- SEO & AI Copywriting – Use SurferSEO to optimize content.

- Automated Webinars – Use WebinarJam to convert leads into buyers.

- AI-Driven Ads – Use Meta AI Ads & Google Ads AI to optimize campaigns.

- AI Email Sequences – Use ConvertKit AI to nurture students.

Example:

If you're selling an Embedded Systems Course, you can:

- Create an AI-powered landing page & lead magnet.

- Run AI-driven Facebook Ads.

- Use ConvertKit AI for automated follow-ups.

Action Step: Set up one AI-driven marketing funnel for your course.

Build & Automate Your Online Course

AI-powered LMS platforms and automation tools make it easier than ever to teach, engage, and monetize your expertise as an engineer.

- Choose the right LMS (Teachable, Thinkific, LearnDash, etc.).

- Leverage AI tools for content creation & automation.

- Use AI-driven sales funnels to attract students.

- Scale your course with pricing models & memberships.

Next Steps:

- Pick an LMS platform and start building your course today.

- Identify one AI tool to streamline your content creation.

- Set up an AI-powered marketing system to grow your audience.

********** COMMUNICATE **********

CHAPTER 5

WHY COMMUNICATION IS THE KEY TO GROWTH

5.1 The Role of Communication in Career Success

In today's fast-changing world, technical expertise alone is not enough to guarantee career success. As an engineer, your ability to communicate effectively plays a crucial role in how far you can go in your profession. Whether you are explaining complex concepts, persuading stakeholders, or selling your expertise, strong communication skills set you apart from the competition.

Why Communication is Essential for Engineers

- Bridging the Gap Between Ideas and Execution – Engineers often work on cutting-edge solutions, but without clear communication, even the best ideas may never see the light of day.

- Influencing Decision-Makers – Whether presenting to clients, managers, or investors, effective communication helps you gain trust and approval for your projects.

- Enhancing Collaboration – Engineering is rarely a solo endeavor. The ability to work seamlessly with teams, both technical and non-technical, is essential.

- Accelerating Career Growth – Those who can articulate their ideas well are more likely to be seen as leaders, opening doors to promotions and entrepreneurial success.

The Three Dimensions of Communication: Say, Sell, and Serve

1. Say – Sharing your knowledge and expertise with the world through various channels such as social media, public speaking, blogs, and networking.

2. Sell – Convincing others of the value of your expertise, whether it's through marketing, client negotiations, or business sales.

3. Serve – Providing ongoing value to your clients and audience through effective follow-ups, support, and community building.

AI's Role in Transforming Communication for Engineers

With AI-powered tools, engineers can enhance their communication through:

- Automated content creation for blogs, emails, and social media.

- AI-driven customer engagement through chatbots and personalized messaging.

- Data-driven storytelling to present technical insights in an engaging manner.

Mastering communication is no longer optional—it's a core skill that engineers must develop to achieve long-term success in their careers.

In the following chapters, we'll dive deeper into how you can master the art of Saying, Selling, and Serving to unlock new career and business opportunities.

5.2. How Engineers Can Leverage Communication for Influence, Income, and Impact

Engineering is not just about solving technical problems—it's about influencing industries, driving innovation, and creating real-world impact.

However, even the most groundbreaking ideas can go unnoticed if they are not effectively communicated. Mastering communication is the key to positioning yourself as a leader, monetizing your expertise, and making a meaningful difference in the world.

1. Communication for Influence: Establishing Authority & Thought Leadership

Engineers who communicate effectively can establish themselves as experts in their field. Influence comes from consistently sharing valuable insights, engaging in discussions, and positioning oneself as a go-to authority.

Ways to build influence through communication:

- Public Speaking & Presentations – Speak at conferences, webinars, and industry events.

- Writing & Content Creation – Publish blogs, whitepapers, or LinkedIn articles.

- Social Media & Personal Branding – Engage with your audience by sharing thoughts on emerging technologies and industry trends.

- Networking & Community Engagement – Join professional groups, forums, and mentorship programs.

Impact: Influence leads to recognition, partnerships, and leadership opportunities in your industry.

2. Communication for Income: Monetizing Your Expertise

Your ability to communicate effectively can directly translate into financial growth. Engineers who can articulate the value of their knowledge can turn their expertise into profitable ventures.

Ways to use communication for income generation:

- Selling Your Expertise – Package your knowledge into courses, workshops, consulting, or e-books.

- Persuasive Storytelling – Sell products or services by effectively presenting their benefits.

- Negotiation & Value-Based Selling – Whether pitching a startup or negotiating a job offer, strong communication skills can significantly impact earnings.

- AI-Driven Marketing & Sales – Utilize AI tools for automated client outreach, personalized content, and conversion-driven strategies.

Impact: Monetization allows engineers to achieve financial freedom while doing what they love.

3. Communication for Impact: Driving Change & Innovation

Great engineers don't just build solutions—they inspire and empower others. Communication is the bridge between knowledge and meaningful societal impact.

Ways to use communication for impact:

- Teaching & Mentorship – Share knowledge to uplift the next generation of engineers.

- Advocacy & Thought Leadership – Speak up on issues like sustainability, ethical AI, and social impact.

- Creating Scalable Solutions – Use digital platforms to spread knowledge on a global scale.

- Building Collaborative Ecosystems – Foster networks that lead to innovation, cross-disciplinary projects, and industry advancements.

Impact: A well-communicated vision can inspire action, drive change, and leave a lasting legacy.

Engineers as Communicators of the Future

In the AI era, engineers must embrace communication as a tool for career success, business growth, and societal transformation. Whether you want to be an industry influencer, increase your income, or make a lasting impact, mastering communication will set you apart.

5.3. The Difference Between Expertise and Perceived Expertise

In today's fast-paced world, engineers must not only be experts but also be seen as experts to unlock career growth, influence, and financial success. There is a significant difference between expertise (your actual skills and knowledge) and perceived expertise (how others recognize and value your expertise). Understanding this distinction and bridging the gap between the two is crucial for engineers aiming to establish themselves as industry leaders.

1. Expertise: The Foundation of Real Knowledge and Skill

Expertise is your actual depth of knowledge, experience, and ability in a particular domain. It comes from years of education, practice, experimentation, and real-world problem-solving.

How real expertise is built:

- Deep technical knowledge gained through formal education and continuous learning.

- Hands-on experience in solving real-world engineering challenges.

- Problem-solving and innovation within specialized fields.

- Ability to teach, mentor, and explain complex topics simply.

- Staying updated with the latest trends, technologies, and advancements.

Example: An engineer with 10+ years of experience in AI-driven automation, who has worked on real-world projects, optimized systems, and solved complex challenges.

Reality Check: Many engineers are highly skilled but remain unnoticed because they do not communicate their expertise effectively.

2. Perceived Expertise: How the World Recognizes You

Perceived expertise is how others see your knowledge, credibility, and authority. It is shaped by how well you communicate, market, and position yourself in your industry.

What influences perceived expertise?

- Online Presence – Do you share insights, publish content, or speak at events?

- Branding & Visibility – Do people recognize your name as an expert?

- Social Proof – Are you endorsed by industry leaders, media, or testimonials?

- Storytelling & Influence – Can you communicate complex ideas in a way that captivates and educates?

Example: An engineer who regularly shares AI insights on LinkedIn, writes articles, speaks at conferences, and engages with the tech community will be perceived as an expert, even if they have less hands-on experience than someone who works behind the scenes.

Reality Check: Many engineers with lesser expertise are seen as industry leaders because they actively share their knowledge and insights.

3. The Gap: Why Real Experts Often Stay Invisible

Many engineers assume that expertise alone will lead to recognition, but in reality, the world values visibility and perception just as much as skill.

Problem:

- You may have deep expertise, but if no one knows about it, you are limiting your opportunities.

- Many less skilled engineers are perceived as experts simply because they communicate and market themselves better.

Solution:

- Engineers must learn to position themselves as experts by communicating effectively.

- Build a strong personal brand through articles, videos, talks, and social engagement.

- Master persuasive storytelling to explain complex ideas in an engaging way.

4. Bridging the Gap: Turning Real Expertise into Recognized Authority

To succeed in today's AI-driven world, engineers must combine their real expertise with strategies that enhance perceived expertise.

How to increase perceived expertise without compromising authenticity:

- Document Your Journey – Share case studies, lessons, and industry insights.

- Leverage Digital Platforms – Write blogs, create videos, or participate in podcasts.

- Speak & Present – Join tech panels, webinars, and conferences.

- Show Social Proof – Publish testimonials, endorsements, or collaborations.

- Educate Others – Conduct workshops, mentor young engineers, and create learning content.

Example: If you are an expert in renewable energy, start sharing case studies, industry reports, and best practices. Over time, people will see you as a thought leader in the field.

Expertise Alone Is Not Enough

To maximize your career growth, income, and impact as an engineer, you must balance both real expertise and perceived expertise. In the AI era, being skilled is not enough—you must also be known for your skills.

CHAPTER 6

COMMUNICATE TO SAY – ABOUT YOU AND YOUR VALUE

6.1. Social Media Strategies for Engineers

As an engineer, social media isn't just for entertainment—it's a powerful tool to establish your authority, showcase your expertise, and attract opportunities. Whether you want to share insights, sell your services, or grow your personal brand, the right social media strategy can help you stand out in your niche.

In this section, we'll cover how to choose the right platform, create engaging content, grow your audience, and leverage AI for automation.

1. Choosing the Right Social Media Platform

Not all platforms are equal. Each one serves a different purpose and audience. Engineers must focus on the right platform based on their niche and goals.

Best Social Media Platforms for Engineers

- LinkedIn – Best for professional branding, networking, and sharing insights.

- YouTube – Best for tutorials, product demos, and technical deep dives.

- Twitter/X – Best for quick insights, industry news, and engaging discussions.

- Instagram & TikTok – Best for visual content, behind-the-scenes, and short explainers.

- Reddit & Quora – Best for answering niche technical questions and positioning yourself as an expert.

Example:

- A Mechanical Engineer showcasing 3D CAD models → LinkedIn & YouTube.

- An AI Engineer discussing the latest tech → Twitter & LinkedIn.

- A Freelance Engineer attracting clients → LinkedIn, Reddit, & Quora.

Action Step: Choose 1-2 platforms based on your expertise and audience.

2. Crafting Your Social Media Strategy

Step 1: Define Your Personal Brand

Your online presence should clearly communicate:

- Who you are (Your expertise)

- What you offer (Your niche)

- Who you help (Your audience)

Step 2: Create a Content Pillar Strategy

Don't post randomly—stick to content pillars that align with your expertise.

Content Pillars for Engineers

- Educational Posts – Tutorials, explainer videos, industry trends.

- Experience-Based Posts – Lessons from your projects, case studies.

- Engagement Posts – Polls, questions, and discussions.

- Behind-the-Scenes – Your daily work, tools, and processes.

- Success Stories – Client testimonials, project showcases.

Example:

- A Civil Engineer can post before/after project photos + a time-lapse.

- A Software Engineer can share code snippets + AI-powered automation tips.

Action Step: Choose 3-4 content pillars to focus on.

3. Creating High-Value Content

LinkedIn: Your Professional Thought Leadership Hub

- Write engaging text posts (Stories, case studies, insights).

- Share carousel posts with step-by-step technical breakdowns.

- Post short videos of your projects or solutions.

- Engage with other professionals in comments and groups.

Example:

A Structural Engineer can post:

- "5 Lessons I Learned from Designing a Skyscraper in Dubai" (Personal insight).

- "How AI is Helping Engineers Detect Structural Failures Faster" (Tech trends).

 Action Step: Post twice a week with a mix of insights, visuals, and engagement posts.

YouTube: The Best Platform for Engineers

- Create tutorial videos on tools & software.

- Showcase your projects (Before/After walkthroughs).

- Do live Q&A sessions to engage with your audience.

- Leverage AI video editors (Synthesia, Descript) to automate content.

Example:

An AI Engineer can make:

- "How to Use Python for AI-Powered Data Analysis" (Tutorial).

- "The Future of AI in Mechanical Engineering" (Industry Trends).

Action Step: Start with one high-quality video per week.

Twitter/X: The Best for Quick Insights

- Share engineering tips in a Twitter thread.

- Engage in industry conversations.

- Share AI-generated infographics.

Example:

A Cybersecurity Engineer can tweet:

"10 Cybersecurity Mistakes Engineers Make & How to Avoid Them 📜 :"

Action Step: Post 1-2 tweets per day with valuable insights.

Instagram & TikTok: The Best for Visual Content

- Post time-lapse videos of your engineering work.

- Use AI tools (Runway ML, Synthesia) to create short-form videos.

- Showcase tools, workspaces, and processes.

Example:

An Electrical Engineer can post:

"Watch this circuit board assembly in 60 seconds! ⚡ " (Time-lapse video).

Action Step: Post 2-3 short videos per week.

4. Growing Your Audience & Engagement

The 3E Formula for Social Media Growth

- Engage: Reply to comments, DMs, and other people's posts.

- Educate: Share actionable insights, tips, and case studies.

- Expand: Collaborate with other professionals & engineers.

Growth Hacks for Engineers

- Use AI-generated captions & hashtags (Flick, Copy.ai).

- Repurpose long-form content into tweets, reels, and carousels.

- Join and contribute to niche engineering groups.

Example:

A Robotics Engineer can:

- Post a YouTube tutorial.

- Share a short clip on LinkedIn.

- Convert it into a Twitter thread.

Action Step: Engage with at least 10 relevant posts daily to increase visibility.

5. Leveraging AI for Social Media Automation

Don't spend hours creating content—use AI tools to automate, optimize, and grow faster.

Best AI Tools for Engineers on Social Media

- ChatGPT / Jasper AI – AI-generated post ideas & captions.

- Canva AI / Adobe Firefly – AI-powered graphic design.

- Descript / Synthesia AI – AI video editing & voiceovers.

- Hootsuite / Buffer – AI-powered post scheduling.

- Flick / Copy.ai – AI-generated hashtags & SEO optimization.

Example:

An Automation Engineer can use Synthesia AI to create:

"5 Ways AI is Transforming Manufacturing" (with AI voiceover & animations).

Action Step: Use one AI tool to automate your social media workflow.

Build Your Authority as an Engineer on Social Media

By strategically using LinkedIn, YouTube, Twitter, and other platforms, you can grow your brand, attract opportunities, and monetize your expertise.

- Choose the right platform (LinkedIn, YouTube, etc.)

- Follow a content pillar strategy (Education, Engagement, Behind-the-Scenes, etc.)

- Use AI tools for automation & efficiency

- Engage with your audience and industry leaders

Next Steps:

- Pick one platform and start posting twice a week.

- Use one AI tool to streamline your workflow.

- Engage with 10 people daily in your industry.

6.2. Websites, Blogs, and Online Branding for Engineers

In today's digital age, having an online presence is not optional—it's essential.

A well-crafted website and blog establish your authority, help you attract opportunities, and position you as an expert in your engineering niche.

Why Engineers Need Online Branding?

- Stand Out in a Crowded Market – Showcase your expertise.

- Attract High-Value Clients & Employers – Build credibility.

- Monetize Your Knowledge – Sell courses, consultations, or products

- Control Your Narrative – Unlike social media, your website is your platform.

1. Creating a High-Impact Website

Your website is your digital headquarters. It should communicate who you are, what you do, and how you can help others.

Essential Pages for an Engineer's Website

- Homepage – First impression matters. Clearly state your expertise & value.

- About Page – Share your journey, experience, and what makes you unique.

- Services/Offerings – Showcase what you do (Consulting, Training, Digital Products, etc.).

- Portfolio/Case Studies – Display your work, projects, and success stories.

- Blog – Educate and attract visitors through valuable content.

- Contact Page – Make it easy for people to reach out.

Example:

A Mechanical Engineer can have:

- Homepage: "Helping Businesses Optimize Manufacturing Processes with AI"

- Portfolio: Case studies of past projects.

- Blog: Articles on Industry Trends, Case Studies, and How-To Guides.

Action Step: Start with a simple website using WordPress, Wix, or Webflow.

2. Building a Powerful Blog to Establish Authority

A blog is a content powerhouse. It helps you attract the right audience, rank on Google, and showcase your expertise.

What to Write About?

- Educational Posts – Tutorials, how-to guides, and explanations.

- Industry Insights – The latest trends in your field.

- Case Studies – Real-world applications of your work.

- Opinion Pieces – Your take on AI, emerging technologies, and best practices.

Example Topics for Engineers:

- Software Engineer: "How AI is Automating Code Review & Testing"

- Civil Engineer: "5 Ways AI is Improving Construction Efficiency"

- Electrical Engineer: "How Smart Grids Are Revolutionizing Energy Distribution"

Action Step: Write one blog post per week to build your online presence.

3. Strengthening Your Online Branding

Your brand is your reputation. The more consistent and professional your online presence, the more trust you build.

Key Branding Elements for Engineers

- Professional Logo & Design – Keep it clean and minimal.

- Consistent Color & Font Scheme – Create a recognizable identity.

- Personalized Domain Name – Example: www.YourNameEngineer.com

- Optimized SEO Content – Ensure your content ranks on Google.

- Clear Value Proposition – Define your niche and who you help.

Example Branding Statement:

"I help mechanical engineers integrate AI-driven automation to improve efficiency & reduce costs."

Action Step: Define your niche, colors, fonts, and messaging for consistency.

4. Leveraging AI for Website & Blog Growth

AI tools make it easier to build, manage, and optimize your online presence.

Best AI Tools for Engineers' Websites & Blogs

- ChatGPT / Jasper AI – Generate blog ideas & write articles faster.

- Surfer SEO / NeuronWriter – Optimize blogs for search rankings.

- Framer AI / Wix ADI – AI-powered website design & development.

- Grammarly / Hemingway Editor – Refine blog content & improve readability.

- Canva / MidJourney AI – Create high-quality visuals for branding.

Example:

An Automation Engineer can use Jasper AI to create a blog on "How AI is Transforming Manufacturing" in minutes.

Action Step: Use one AI tool to speed up your content creation.

Build Your Digital Footprint as an Engineer

By building a website, starting a blog, and creating a strong online brand, you can:

- Attract new opportunities & clients

- Showcase your expertise with authority

- Monetize your knowledge through digital products

Next Steps:

- Set up a simple website (Wix, WordPress, Webflow)

- Write one blog post per week in your engineering niche

- Use AI tools to automate and optimize your content

6.3. AI-Powered Content Marketing and SEO for Engineers

In today's digital world, content marketing and SEO (Search Engine Optimization) are essential for engineers looking to establish expertise, attract clients, and grow their careers or businesses. AI is revolutionizing both, making it easier to create, optimize, and distribute high-value content.

This chapter will guide you through how engineers can use AI-powered tools to improve content marketing, boost SEO, and reach the right audience effectively.

Why Engineers Need AI-Driven Content Marketing?

- Position Yourself as an Industry Expert – Regular content builds credibility.

- Reach the Right Audience Faster – AI tools optimize your content for discoverability.

- Automate and Save Time – AI can generate blog posts, optimize SEO, and analyze trends.

- Increase Website Traffic and Conversions – More visibility = more clients & opportunities.

Whether you're an engineering consultant, course creator, or product developer, AI-powered marketing gives you an edge over competitors.

AI-Powered Content Creation for Engineers

AI tools help generate high-quality, engaging content with minimal effort. Here's how you can leverage them:

Step 1: AI for Content Research & Idea Generation

Before writing, you need the right topics. AI tools can analyze search trends, competitors, and industry news to suggest high-impact topics.

Best AI Tools:

- ChatGPT / Jasper AI – Generate blog ideas & outlines.

- AnswerThePublic – Find trending questions in your niche.

- BuzzSumo – Analyze popular engineering-related content.

- Google Trends – Identify what's trending in engineering fields.

Example:

A Civil Engineer researching "Smart Cities & AI" can use AnswerThePublic to find questions like:

- How is AI being used in smart city planning?

- What are the latest AI-driven urban infrastructure projects?

Step 2: AI for Writing & Content Generation

Once you have a topic, AI can help write blog posts, LinkedIn articles, and social media content.

Best AI Writing Tools:

- Jasper AI / ChatGPT – Draft high-quality blogs & articles.

- Copy.ai / Writesonic – Generate marketing copy & landing pages.

- QuillBot – Rewrite & enhance content readability.

- Hemingway Editor – Improve clarity & conciseness.

Example:

A Software Engineer writing about "AI in Code Optimization" can use ChatGPT to generate:

- A structured blog post outline.

- Key points for LinkedIn posts.

- Engaging social media captions.

Action Step: Write your next blog post using AI and refine it for authenticity.

AI-Powered SEO: How Engineers Can Rank Higher on Google

SEO ensures that your content reaches the right audience. AI simplifies SEO by analyzing keywords, competition, and ranking strategies.

Step 1: AI Keyword Research

AI can identify the best keywords for your niche, helping you rank on Google.

Best AI SEO Tools:

- Surfer SEO – Find & optimize content with high-ranking keywords.

- Ahrefs / SEMrush – Analyze competitor keywords.

- Frase.io – AI-driven content optimization.

- NeuronWriter – AI-powered content suggestions based on ranking pages.

Example:

An Automation Engineer creating a blog on "AI in Industrial Automation" can use Ahrefs to find the best-ranking keywords:

- "AI-driven automation in manufacturing"

- "How AI is optimizing industrial processes"

Action Step: Use Surfer SEO or Ahrefs to research keywords before writing.

Step 2: AI-Powered Content Optimization

AI can analyze top-ranking pages and suggest ways to improve your content for better visibility.

Best AI Content Optimization Tools:

- Yoast SEO (WordPress Plugin) – Optimize blog posts for SEO.

- Grammarly + Surfer SEO – Enhance readability & keyword integration.

- NeuronWriter – AI-powered content analysis based on competitors.

Example:

A Mechanical Engineer writing about "AI in Manufacturing" can use Surfer SEO to:

- Optimize headings & keywords.

- Add missing subtopics from top-ranking articles.

- Improve readability for SEO ranking.

Action Step: Run your existing content through Surfer SEO to check for improvements.

AI-Driven Content Distribution: Reaching More Engineers & Clients

Creating content isn't enough—you need to promote it effectively. AI tools can automate content distribution across platforms.

Step 1: AI for Social Media Marketing

AI tools schedule, optimize, and even create social media posts for engineers.

Best AI Social Media Tools:

- Buffer / Hootsuite – Schedule & automate posts.

- Lumen5 / Pictory AI – Convert blog posts into engaging videos.

- Canva AI – Generate social media visuals.

- ChatGPT for LinkedIn – Generate engaging post ideas.

Example:

A Data Engineer who writes a blog on "AI in Data Analytics" can use Lumen5 to turn it into a short video for LinkedIn & YouTube.

Action Step: Automate weekly social media posts using Buffer or Hootsuite.

Step 2: AI for Email Marketing

Email marketing is a high-converting strategy for engineers offering courses, consulting, or digital products.

Best AI Email Marketing Tools:

- ConvertKit / Mailchimp AI – Automated email sequences.

- ChatGPT for Emails – Generate personalized email copy.

- Grammarly AI – Refine email writing for clarity.

Example:

An Embedded Systems Engineer launching an online course can use ConvertKit AI to:

- Automate welcome emails & course promotions.

- Personalize emails based on reader behavior.

- Increase engagement with AI-optimized subject lines.

Action Step: Set up a lead magnet & email automation using ConvertKit.

AI is the Future of Content Marketing & SEO for Engineers

By leveraging AI for content creation, SEO, and marketing, engineers can:

- Create high-value content faster

- Rank higher on Google with AI-powered SEO

- Attract clients, employers, and opportunities

- Automate content distribution for maximum reach

Next Steps:

- Use AI tools to generate & optimize content.

- Automate social media & email marketing.

- Improve SEO with AI-powered research & analysis.

6.4. Online & Offline Networking Strategies for Engineers

Networking is one of the most powerful tools for career growth. As an engineer, your ability to connect with the right people, share your expertise, and build relationships can open doors to job opportunities, business collaborations, and personal growth.

This chapter will cover how engineers can leverage both online and offline networking strategies to grow their influence, find opportunities, and establish credibility in their field.

Why Engineers Need Strong Networking Skills?

- Job & Business Opportunities – Most jobs and projects come through referrals.

- Learning & Growth – Networking exposes you to new technologies and trends.

- Building Your Personal Brand – Visibility helps you stand out in your industry.

- Collaboration & Innovation – Great ideas and partnerships emerge from strong networks.

Part 1: Online Networking for Engineers

The internet has made networking easier and more powerful than ever before. Here's how you can connect with industry leaders, potential clients, and collaborators without leaving your desk.

1. LinkedIn – The #1 Platform for Engineers

LinkedIn is a must-have for engineers looking to build their professional network. Here's how to optimize it:

Optimize Your LinkedIn Profile:

- Professional Profile Photo – A clear, high-quality headshot.

- Compelling Headline – Describe your expertise (e.g., "AI & Robotics Engineer | Helping Industries Automate with Smart Solutions").

- Engaging About Section – Share your story, skills, and what you offer.

- Experience & Projects – Showcase technical expertise and achievements.

How to Network on LinkedIn:

- Engage with Posts – Like, comment, and share insights.

- Post Valuable Content – Share engineering insights, case studies, and lessons.

- Send Personalized Connection Requests – Mention mutual interests when connecting.

- Join LinkedIn Groups – Engage in discussions related to your expertise.

Action Step: Connect with 5 industry professionals and comment on 3 posts daily.

2. Twitter & X – Engaging in Engineering Discussions

Twitter (X) is fast-paced but great for staying updated and engaging with industry leaders.

How to Network on Twitter:

- Follow experts, companies, and influencers in your field.

- Share engineering insights and industry news.

- Engage in relevant threads and discussions.

- Use hashtags like #Engineering, #TechInnovation, and #AI.

Example: A Mechanical Engineer could follow and engage with Elon Musk, ASME, or AutoCAD influencers.

Action Step: Tweet or reply to an industry post once a day to stay active.

3. Engineering Forums & Online Communities

Joining niche engineering communities can help you connect with like-minded professionals.

- Best Online Communities for Engineers:

- Reddit Engineering Forums – r/engineering, r/electricalengineering, r/civilengineering.

- GitHub & Stack Overflow – Great for software and hardware engineers.

- Quora & Medium – Answer engineering-related questions to establish authority.

- Industry-Specific Groups – AI forums, IEEE communities, or technical Slack groups.

Action Step: Join one online engineering forum and answer questions weekly.

4. Online Events, Webinars & Virtual Meetups

With AI-driven tools, engineers can attend conferences, webinars, and workshops from anywhere.

How to Find & Leverage Online Events:

- Follow industry leaders and companies for event announcements.

- Register for relevant online webinars, tech talks, and virtual summits.

- Participate in discussions and ask insightful questions.

- Connect with speakers and attendees on LinkedIn.

Example: A Cybersecurity Engineer attending a DEFCON virtual meetup can network with global experts.

Action Step: Attend one online engineering event per month and connect with speakers.

Part 2: Offline Networking for Engineers

While online networking is powerful, nothing beats face-to-face connections. Here's how to maximize in-person interactions for career growth.

1. Industry Conferences & Technical Expos

Attending conferences, expos, and seminars is a great way to meet industry leaders and potential clients.

How to Network at Conferences:

- Research attendees and speakers before the event.

- Prepare a short elevator pitch about your expertise.

- Carry professional business cards or a digital contact-sharing app.

- Attend breakout sessions and panel discussions.

- Follow up with connections after the event on LinkedIn.

Example: A Renewable Energy Engineer at a solar power expo can meet potential employers, investors, and collaborators.

Action Step: Attend at least one industry conference per year.

2. Professional Associations & Meetups

Joining engineering societies and meetup groups expands your professional network.

Top Engineering Associations to Join

- IEEE (Institute of Electrical and Electronics Engineers)

- ASME (American Society of Mechanical Engineers)

- NSPE (National Society of Professional Engineers)

- Local Engineering Chambers & Meetup Groups

Action Step: Become a member of a professional association in your field.

3. University Alumni Networks & Engineering Clubs

Your college alumni network is a goldmine for networking!

How to Leverage Alumni Networks:

- Attend university networking events and engineering club meetings.

- Connect with senior alumni who are in leadership roles.

- Offer mentorship to junior engineers and students.

- Join WhatsApp or Telegram groups of your alumni network.

Example: A Structural Engineer from IIT Madras can use the IIT Alumni Network to connect with top construction firms.

Action Step: Reach out to two alumni members in your field.

4. Engineering Hackathons & Competitions

If you're in software, robotics, or automation, hackathons and competitions are excellent networking spots.

Best Engineering Hackathons & Competitions:

- NASA Space Apps Challenge – Great for aerospace & AI engineers.

- Google Code Jam / Kaggle – For software and AI engineers.

- Shell Eco-Marathon – For mechanical and automotive engineers.

Action Step: Participate in one engineering competition per year.

5. Build Your Engineer's Network Strategically

Networking isn't just about meeting people—it's about building meaningful relationships. Whether online or offline, being consistent, helpful, and valuable in your interactions will create strong professional opportunities.

- Online Networking – Optimize LinkedIn, engage in forums, attend webinars.

- Offline Networking – Attend conferences, join professional groups, meet alumni.

Next Steps:

- Optimize your LinkedIn profile today.

- Join an engineering group or forum this week.

- Attend a live networking event in the next month.

CHAPTER 7

COMMUNICATE TO SELL –CONVERT YOUR VALUE INTO SALES

7.1. Value-Based Sales Strategies for Engineers

As an engineer, selling might not come naturally to you. You're used to designing solutions, solving problems, and working with technical details—not persuading people to buy something. But here's the truth: Every successful engineer is also a great salesperson.

Whether you're selling a product, service, or even yourself (as an expert), the key is value-based selling—focusing on how your expertise solves real-world problems and delivers tangible benefits.

This chapter will guide you through effective sales strategies tailored for engineers, so you can confidently sell without feeling like a salesperson.

Why Engineers Must Learn to Sell?

1. Better Career & Business Growth – Selling your expertise helps you get promotions, projects, and clients.

2. Increased Trust & Credibility – When you sell based on value, people see you as a problem-solver, not just a salesperson.

3. Higher Income Potential – The best-paid engineers know how to position and monetize their knowledge.

4. Stronger Client Relationships – Value-based sales create long-term trust and repeat business.

What is Value-Based Selling?

Traditional selling focuses on features (what a product does).

Value-based selling focuses on benefits & impact (how it solves a problem).

Example: Instead of saying:

"Our AI software processes data at 10x speed."

Say this:

"With our AI software, you can analyze data in minutes instead of hours—saving you time and boosting productivity."

The key: Sell outcomes, results, and transformation—not just technology.

1. Understanding Customer Psychology

Engineers often focus on logic, but buying decisions are mostly emotional.

People buy when they:

- Trust you.

- See clear benefits.

- Feel that your solution will improve their life or business.

How to connect emotionally?

- Understand their pain points.

- Show empathy—speak their language, not technical jargon.

- Share real-life examples and case studies.

Example: If you're selling an IoT automation system, don't just talk about sensors and connectivity. Talk about how it reduces energy costs and increases efficiency for factory owners

2. The 4-Step Value-Based Sales Framework

1. Identify the Problem

- Ask the right questions to understand the client's pain points.

- Example questions:

   o What challenges are you facing in [your industry]?

   o How is this problem affecting your productivity/revenue?

o Have you tried other solutions? What worked, what didn't?

2. Position Your Solution

- Show how your expertise directly solves their problem.

- Don't overload with technical details—focus on benefits.

- Example: If you're selling an AI-based predictive maintenance system, say:

- "With our AI system, you can prevent machine failures before they happen, reducing downtime by 40% and saving thousands in repairs."

3. Provide Proof & Value

- Clients need reassurance. Provide:

  o Case studies & testimonials.

  o Data-driven results.

  o Comparisons with other solutions.

4. Close with Confidence

- Instead of "pushing" for a sale, guide them to a decision.

- Ask questions like:

  o "Would reducing downtime by 40% make a big impact for your business?"

- o "What would achieving these results mean for your company?"

- When they see the value clearly, the decision becomes easy

3. Practical Sales Techniques for Engineers

1. Use the 'Consultative Selling' Approach

Instead of pitching, become a trusted advisor.

- Listen more than you talk.

- Diagnose their problem before offering a solution.

- Give free value (tips, insights, a small free trial).

2. Tell Stories Instead of Giving Data

People remember stories, not statistics.

- Use case studies:

  - o Example: One of our clients reduced energy costs by 25% with our automation system.

- Share customer success stories.

3. Price Based on Value, Not Just Cost

Instead of saying:

- "Our service costs $500."

- Say:

- "For a $500 investment, you'll save $5,000 in operational costs this year."

- The focus should always be on Return on Investment (ROI).

4. Using AI Tools to Enhance Sales

AI is revolutionizing sales for engineers. Here's how:

- AI Chatbots – Automate customer queries (Drift, ChatGPT, ManyChat).

- AI-Powered CRMs – Manage leads and follow-ups (HubSpot, Salesforce).

- AI-Based Sales Predictions – Forecast trends & customer needs (Gong.io).

Sell with Confidence, Deliver with Impact

Selling isn't about convincing—it's about helping.

- Focus on value and transformation, not just technology.

- Understand customer psychology and emotions.

- Use consultative selling and storytelling techniques.

- Leverage AI tools to enhance efficiency.

Next Steps:

- Identify a client or employer challenge and craft a value-based sales pitch.

- Practice explaining your expertise in benefits, not features.

7.2. Understanding Customer Psychology & Trust-Building

As an engineer, you're used to working with facts, data, and logic.

But when it comes to business, sales, and career growth, decisions are driven by psychology and trust—not just technical superiority.

Whether you're selling a product, service, or even yourself (as an expert), understanding how customers think, what drives their decisions, and how to build long-term trust is essential.

Let's break it down step by step.

Why Customer Psychology Matters for Engineers

Most engineers believe:

- "If I build a great product or provide a great service, customers will automatically buy."

But in reality:

- Customers don't just buy the best product; they buy from people they trust and connect with.

- Even the most technically perfect solution will fail in the market if it doesn't align with the customer's emotions, beliefs, and needs.

The Key Insight:

- People make buying decisions emotionally first, then justify them logically.

Example: A client considering your AI-based automation tool won't just think about its technical features. They'll think about:

- "Will this make my job easier?"

- "Can I trust this person/company?"

- "What if something goes wrong?"

Your job is to address these concerns and build trust.

1. The 3 Key Psychological Triggers in Decision-Making

1. Pain & Problems – People Pay to Solve Issues

- Customers act faster when they feel a pain point.

- Your expertise should clearly solve a major problem.

Example: Instead of saying:

- "This software speeds up data processing." (Too generic)

Say:

- "Are you losing hours manually analyzing data? Our software automates it, so you can focus on high-value tasks." (Problem-focused)

2. Trust – People Buy from Who They Believe In

- Expertise alone is not enough—people need to trust you.

Ways to Build Trust:

- Authenticity – Be honest about what you offer.

- Consistency – Deliver quality every time.

- Transparency – Show proof of your work (case studies, testimonials).

- Social Proof – If others trust you, new customers will too.

Example: Amazon reviews influence buying decisions because they show proof that others trust the product.

3. Value – People Buy Based on ROI (Return on Investment)

- Customers don't care about what you do—they care about what they get.

Formula for Value Communication:

- Outcome + Timeframe + Emotion

Example: Instead of saying:

- "This AI system improves workflow efficiency." (Too vague)

Say:

- "With our AI system, you'll complete projects 50% faster, reduce stress, and boost revenue within 3 months." (Clear, time-bound, and outcome-driven)

2. How to Build Long-Term Trust

Trust is not built in a single conversation—it's built over time through actions.

Step 1: Understand Their Needs Deeply

- Ask more questions than you answer.

- What are their biggest challenges?

- What are their long-term goals?

- What solutions have they tried before?

Example: If you're offering an IoT-based monitoring system for factories, don't just say:

- "This system tracks machine performance."

Instead, ask:

- "How much does downtime cost you per year? What if you could reduce that by 30%?"

Step 2: Educate, Don't Just Sell

- Position yourself as a trusted advisor, not just a seller.

- Share insights, case studies, and helpful content.

- Provide value upfront—offer free tips, reports, or webinars.

- Explain why a solution works, not just what it does.

Example: If you're selling AI-driven cybersecurity solutions, don't just say:

- "This tool prevents data breaches."

Instead, educate:

- "80% of cyber-attacks exploit human error. Our AI system analyzes behavior patterns and prevents breaches before they happen."

Step 3: Be Consistent & Reliable

- The fastest way to break trust is by being inconsistent.

- Respond to inquiries on time.

- Deliver what you promise.

- Stay engaged with clients even after they buy.

Example: If you provide an online course on engineering automation, don't disappear after the sale. Follow up with:

- Additional resources.

- Live Q&A sessions.

- Personalized support.

Step 4: Use Social Proof & Testimonials

- People trust other people's experiences.

- Case studies.

- Video testimonials.

- Client success stories.

Example: Instead of saying:

- "This AI tool improves productivity."

Show proof:

- "After using our AI tool, XYZ Company reduced errors by 40% and increased efficiency by 25% in just 3 months."

3. How AI Can Help in Understanding Customer Psychology

AI can analyze customer behavior, preferences, and engagement patterns, making trust-building easier.

- AI Chatbots & Virtual Assistants – Provide 24/7 support & answer customer questions instantly.

- AI Sentiment Analysis – Understand customer emotions through feedback & reviews.

- AI-Powered CRMs – Track customer interactions & personalize communication.

Example: AI tools like HubSpot and Salesforce can predict what a customer needs before they even ask for it.

Trust is the Foundation of Growth

- Customers don't buy products—they buy solutions and trust.

- Focus on pain points, emotions, and ROI.

- Educate, engage, and deliver consistent value.

- Use AI tools to enhance customer insights & trust-building.

Next Steps:

- Identify a real-world customer problem in your niche.

- Develop a value-based pitch that connects with their emotions.

- Find ways to build trust (testimonials, free insights, case studies).

7.3. Persuasive Communication & Storytelling for Engineers

As an engineer, you might be used to explaining things logically—with facts, numbers, and data. But when it comes to influencing people, selling ideas, or getting buy-in, logic alone is not enough.

People don't just buy into products or services—they buy into emotions, experiences, and stories.

Whether you're pitching a project, selling a product, or marketing your expertise, mastering persuasive communication and storytelling is key to success.

1. Why Engineers Need Persuasive Communication?

Persuasive communication = Making people say YES with confidence.

- If you can persuade effectively, you can:
  - Sell your ideas to bosses, clients, or investors.
  - Convince customers to buy your solutions.
  - Build a strong brand as an expert in your niche.

But here's the mistake many engineers make:

- They overload people with technical jargon and expect them to be convinced.

- Instead, you must communicate in a way that connects emotionally and logically.

2. The Science Behind Persuasion

Persuasion works because of three psychological principles:

1. ETHOS (Credibility) – Why should they trust you?

People buy from experts they trust. You must:

- Show credibility through experience or success stories.

- Be authentic—don't overpromise or exaggerate.

- Provide social proof (testimonials, case studies).

Example:

- "I built an AI tool that predicts machine failures." (Too vague)

- "Over 50 manufacturing companies trust our AI tool to reduce downtime by 30%." (Builds credibility)

2. PATHOS (Emotion) – Why should they care?

People make decisions emotionally first, then justify them logically.

- Use relatable problems and real-life examples.

- Make your message personal and meaningful.

- Show the impact your solution has on their life or business.

Example:

- "This automation tool optimizes workflow efficiency." (Too technical)

- "Ever felt frustrated with repetitive tasks? Our tool frees up your time so you can focus on high-impact work." (Emotion-driven)

3. LOGOS (Logic) – Why does it make sense?

Once you've established trust (ETHOS) and emotion (PATHOS), then use logic (LOGOS).

- Use data, case studies, and clear explanations.

- Show the cost of inaction (how much they'll lose if they don't act).

- Present a clear next step (so they know what to do).

Example:

- "AI automation improves business processes." (Too general)

- "Companies using AI automation see a 25% reduction in errors and save an average of $50,000 per year." (Clear, logical, and valuable)

3. The Power of Storytelling in Persuasion

- FACT: People remember stories 22X more than facts alone.

- STORY = Problem → Struggle → Solution → Success

Here's how you can use storytelling to persuade effectively:

Step 1: Start with a Relatable Problem

- Identify a pain point your audience is facing.

- Describe it in a way they emotionally connect with.

Example:

- "AI helps in cybersecurity." (Too boring)

- "Imagine waking up to find your company's data stolen. That's what happened to a startup that ignored AI security. They lost millions overnight." (Powerful and relatable)

Step 2: Show the Struggle & The Stakes

- Create suspense – What's at risk if they don't take action?

- Highlight common mistakes or misconceptions.

Example:

- "Many companies don't use AI security."

- "They assumed their systems were safe—until hackers found vulnerabilities they never even considered."

Step 3: Introduce the Solution (Your Expertise or Product)

- Explain how your expertise, product, or service solves the problem.

- Make it clear, but not overly technical.

Example:

- "Our AI-based cybersecurity detects threats."

- "Our AI system acts like a 24/7 security guard—identifying and stopping threats before they cause damage."

Step 4: Show the Transformation & Success

- Share a success story (case study, testimonial, or personal experience).

- Show the positive outcome of using your solution.

Example:

- "AI security improves protection."

- "After installing our AI security system, the company blocked 95% of attempted cyberattacks and saved $2M in potential losses."

4. Practical Tips to Make Your Communication More Persuasive

- Use Simple & Clear Language

  - Avoid jargon that confuses people.

  - Explain like you're talking to a smart 12-year-old.

- Speak to Their Pain & Goals

  - Don't talk only about features.

  - Talk about problems solved & benefits gained.

- Use Visuals & Demonstrations

  - Show, don't just tell.

  - Use diagrams, videos, or live examples to make complex ideas easy.

- Repeat Key Messages for Impact

    o Repetition strengthens memory.

    o Reinforce the main benefit multiple times in different ways.

- End with a Clear Call to Action (CTA)

    o Tell them what to do next (Book a call, sign up, try the demo).

5. AI Tools for Persuasive Communication & Storytelling

AI can enhance persuasion & storytelling by:

- Generating compelling marketing copy (ChatGPT, Jasper AI)

- Analyzing customer sentiment to personalize messages (HubSpot, Salesforce)

- Creating engaging visuals to support storytelling (Canva AI, DALL·E)

Master Persuasion, Accelerate Your Growth

- Engineers need persuasion & storytelling to sell ideas and solutions.

- People buy with emotion first, then justify with logic.

- Storytelling makes communication powerful & memorable.

- Use AI tools to enhance persuasion & marketing strategies.

Now, it's your turn!

7.4. AI Tools for Sales Automation & Customer Engagement

1. Why AI is a Game-Changer in Sales?

Traditional sales methods rely heavily on manual outreach, repetitive tasks, and gut-feeling decisions. AI, on the other hand, enables:

- Automation of repetitive tasks (emails, follow-ups, scheduling)

- Better customer insights (AI-driven analytics & predictions)

- Personalized communication (chatbots, recommendation engines)

- Faster response times (AI-powered customer support)

- Improved conversions & sales (data-driven decision-making)

2. AI-Powered Sales Automation Tools

AI for Lead Generation & Outreach These tools help find potential customers, send automated emails, and book meetings.

- HubSpot Sales Hub – AI-powered CRM, automated email sequences, and pipeline tracking.

- Apollo.io – AI-driven prospecting, contact sourcing, and email automation.

- Seamless.ai – Uses AI to find high-value leads and build email lists.

How it helps? AI finds and reaches out to potential customers on autopilot.

AI for Personalized Sales Communication AI tools analyze customer behavior and customize sales messages based on their needs.

- ChatGPT & Jasper AI – Generate high-converting sales emails and social media content.

- Lavender.ai – AI-powered email assistant that improves response rates.

- Crystal Knows – AI-based personality insights for tailored sales pitches.

How it helps? AI personalizes every message, increasing conversion rates.

AI for Automated Follow-Ups & Scheduling

- Calendly + AI – Smart scheduling with AI-driven follow-ups.

- Outreach.io – AI-driven sales engagement for automating follow-ups.

- Drift – AI chatbots that schedule calls and send reminders.

How it helps? AI automates and personalizes follow-ups, ensuring no lead is lost.

3. AI for Customer Engagement & Retention

AI Chatbots & Virtual Assistants

- Drift & Intercom – AI chatbots for real-time customer support & engagement.

- ChatGPT API – Custom AI assistants for answering customer questions.

- ManyChat – AI-powered chatbot for WhatsApp, Messenger, and Instagram.

How it helps? AI chatbots provide instant support to customers 24/7.

AI for Customer Insights & Predictions

- Gong.io – AI sales coaching tool that analyzes conversations.

- Salesforce Einstein AI – Predicts customer behavior and sales opportunities.

- Zoho CRM AI – AI-powered customer engagement & analytics.

How it helps? AI analyzes customer conversations and buying patterns to improve sales pitches.

AI for Hyper-Personalized Customer Experiences

- Persado – AI-generated emotional and persuasive marketing messages.

- Seventh Sense – AI-driven email delivery optimization for higher engagement.

- Dynamic Yield – AI-powered personalization for websites & e-commerce.

How it helps? AI ensures customers receive exactly what they need, boosting engagement and sales.

4. AI in Action: Real-World Use Cases

- Case 1: AI-Powered Lead Generation

  - A software engineer using Apollo.io and Jasper AI saw a 300% increase in leads.

- Case 2: AI Chatbot for Sales

  - An engineer using Drift AI chatbot on their website saw 40% more course sign-ups.

- Case 3: AI-Based Customer Insights

  - A tech consultant using Gong.io saw a 25% higher deal closing rate.

5. The Future: AI-Driven Sales Will Be the New Norm

AI will dominate sales & customer engagement in the coming years. Engineers who embrace AI-powered sales automation will:

- Reach more customers faster.

- Close more deals with less effort.

- Deliver personalized experiences at scale.

6. Take Action: Use AI to Scale Your Sales Efforts

- Step 1 – Start with an AI-powered CRM (HubSpot, Zoho, Salesforce).

- Step 2 – Automate your outreach & follow-ups (Apollo.io, Outreach.io).

- Step 3 – Use AI chatbots to engage with potential customers (Drift, Intercom).

- Step 4 – Leverage AI-generated sales content (Jasper AI, ChatGPT).

- Step 5 – Track customer insights and improve (Gong.io, Persado).

CHAPTER 8

COMMUNICATE TO SERVE – DELIVER VALUE & RETAINING CLIENTS

8.1. Post-Sale Support & Follow-Up Strategies for Engineers

Selling is just the beginning—the real success lies in serving your customers after the sale. If you want to build long-term relationships, increase customer satisfaction, and generate repeat business, post-sale support is a must. Most engineers focus on creating and selling their solutions, but the key to sustainable success is how well you serve after the sale. Strong post-sale support builds trust, loyalty, and referrals, which leads to higher lifetime value (LTV) per customer.

1. Why Is Post-Sale Support Crucial?

- Increases Customer Retention – Happy customers stay longer and buy more.

- Builds Trust & Reputation – Good support makes your brand reliable.

- Boosts Word-of-Mouth Marketing – Satisfied customers refer others.

- Leads to Upselling & Cross-Selling Opportunities – More revenue per customer.

- Reduces Refunds & Complaints – Great support prevents dissatisfaction.

2. Key Post-Sale Support & Follow-Up Strategies

A. Proactive Follow-Up After the Sale

The first 7 days after a sale are crucial—reach out early!

Follow-up emails/messages should:

- Thank the customer for purchasing.

- Provide clear instructions on product/service usage.

- Offer additional resources (guides, FAQs, video tutorials).

- Ask if they need any help or clarification.

Tools to Automate Follow-Ups:

- HubSpot & Zoho CRM – Automate post-sale emails.

- Mailchimp & ConvertKit – Set up email sequences.

- WhatsApp Business API – Automate customer check-ins.

Example Email Sequence:

- Day 1: "Welcome & Thank You!"

- Day 3: "Need Help? Here's a Quick Guide"

- Day 7: "Feedback Request & Next Steps"

B. Provide Exceptional Customer Support

Customers may have questions, issues, or doubts—be ready to help!

Support Channels to Offer:

- Phone Support – Quick problem-solving.

- Live Chat – Instant solutions for customer queries.

- Email Support – Detailed issue resolution.

- AI Chatbots – 24/7 automated customer assistance.

Tools to Enhance Customer Support:

- Zendesk & Freshdesk – AI-powered ticketing & support.

- Drift & Intercom – AI chatbots for instant responses.

- WhatsApp Business & Telegram Bots – Fast, interactive support.

Best Practice: Always respond within 24 hours to maintain customer satisfaction.

C. Offer Value-Added Digital Resources

Engineers love DIY solutions—give them the right tools and guidance.

Content That Adds Value:

- PDF Guides – Step-by-step solutions.

- Video Tutorials – Demonstrations for easy learning.

- Audio Lessons/Podcasts – Learn on the go.

- Exclusive Webinars & Q&A Sessions – Interactive learning.

Best Tools to Share Resources:

- Google Drive & Dropbox – Share PDFs & documents.

- YouTube & Vimeo – Host video tutorials.

- Kajabi, Thinkific & Teachable – Offer exclusive online courses.

- Zoom & Webex – Host live Q&A sessions.

D. Build a Community for Ongoing Support

A strong community keeps customers engaged and loyal.

Ways to Build a Supportive Community:

- Private WhatsApp/Telegram Groups – Quick peer-to-peer support.

- Facebook/LinkedIn Groups – Build a professional network.

- Discord/Slack Channels – Best for technical discussions.

Tools to Manage Communities:

- Facebook Groups & LinkedIn Groups – Professional networking.

- Slack & Discord – Organized discussions.

- Telegram & WhatsApp – Direct engagement with customers.

E. Upselling & Cross-Selling Through Follow-Ups

Selling doesn't stop at the first purchase! The right follow-up can increase revenue per customer.

- Upselling (Premium Upgrade): Offer a more advanced version of the product.

- Cross-Selling (Related Services): Offer complementary products/services.

Example:

- If an engineer buys your "AI & ML for Beginners" course,

- You can offer an "AI for Advanced Engineers" course as an upsell.

AI Tools to Automate Upselling:

- ConvertKit & ActiveCampaign – AI-driven email sequences for upselling.

- ChatGPT & Jasper AI – Personalized messaging for product recommendations.

- Shopify & ClickFunnels – AI-driven upsell pages.

Best Practice: Send personalized recommendations based on their purchase history.

F. Gather Feedback & Improve Your Services

Customer feedback is GOLD—it helps you improve and innovate.

How to Collect Feedback?

- Email Surveys – Send short, engaging feedback forms.

- Google Forms & Typeform – Quick and easy data collection.

- Live Calls – Get personal insights on what customers think.

- AI Sentiment Analysis – Use AI tools to analyze feedback trends.

Best Feedback Collection Tools:

- Google Forms & Typeform – Simple survey creation.

- SurveyMonkey & JotForm – AI-driven customer surveys.

- Trustpilot & Capterra – Collect and showcase reviews.

Example Feedback Email:

- Subject: "We'd love your feedback – Just 1 minute!"

- "Hello [Customer Name], we appreciate you choosing our [Product/Service]. Your feedback helps us improve! Please take 1 minute to share your experience. [Link to survey]"

Bonus Tip: Offer a discount, bonus content, or free consultation as a reward for feedback.

3. Real-World Case Studies of Post-Sale Support Success

- Case 1: AI Chatbot for 24/7 Support

  - An engineer selling automated control systems used Intercom AI chatbot for post-sale customer support.

  - Result: 60% reduction in support requests and faster response time.

- Case 2: Upselling Through Follow-Ups

  - A mechanical engineer offering 3D modeling courses added personalized upsell emails for advanced courses using ConvertKit.

  - Result: 30% increase in sales.

- Case 3: Community-Driven Customer Engagement

  - A software engineer running a Python automation bootcamp created a Slack community for students.

  - Result: 90% customer retention for future courses.

4. Take Action: Build a Post-Sale Strategy for Your Engineering Business

- Automate follow-ups (email sequences, WhatsApp).

- Offer multiple support channels (chatbot, email, live chat).

- Provide digital resources (PDFs, videos, webinars).

- Create a customer community (Telegram, Facebook Groups).

- Gather feedback & improve (Google Forms, surveys).

- Upsell & cross-sell (email automation, personalized offers).

8.2. AI-Powered Chatbots & Automation for Customer Service

In today's digital age, AI-powered chatbots and automation tools are transforming customer service. They help businesses provide instant, efficient, and personalized support 24/7 while reducing workload and costs.

For engineers looking to scale their coaching, courses, or services, AI chatbots and automation can handle FAQs, book appointments, guide users, and even upsell products—all without manual intervention!

1. Why Use AI-Powered Chatbots & Automation?

- 24/7 Availability – AI chatbots never sleep! They provide round-the-clock support.

- Instant Responses – No more waiting. AI handles queries in real time.

- Reduces Workload – Automates repetitive tasks so you can focus on growth.

- Enhances Customer Experience – AI-powered interactions feel personalized.

- Increases Conversions & Sales – AI can recommend products, book demos, and close sales.

2. AI Chatbots vs. Traditional Customer Support

Feature	Traditional Support	AI Chatbots & Automation
Availability	Limited working hours	24/7 real-time responses
Response Time	Can take minutes/hours	Instant replies
Scalability	Expensive to scale	Easily scalable
Consistency	Human errors possible	100% consistent answers
Cost	High (salaries, training)	Lower operational costs
Personalization	Requires manual input	AI-driven customized responses

AI Chatbots don't replace humans, but they handle routine queries so human agents can focus on complex issues.

3. Key Features of AI-Powered Chatbots

- Natural Language Processing (NLP) – Understands human language & context.

- Personalized Responses – AI adapts based on user history & preferences.

- Multichannel Integration – Works across WhatsApp, Facebook, Websites, Emails & more.

- Lead Qualification – Identifies & nurtures potential customers.

- Data Analytics & Insights – Tracks user behavior to improve engagement.

- Seamless Handoff to Humans – Transfers complex queries to human agents.

4. Best AI Chatbots & Automation Tools for Engineers

1. Chatbot Builders for Websites & Social Media

- ChatGPT API – AI-driven, customizable chatbot for advanced automation.

- ManyChat – Best for Facebook Messenger, Instagram, and WhatsApp.

- Tidio – AI-powered chatbot for eCommerce & websites.

- Drift & Intercom – Great for **B2B** businesses (lead generation & support).

2. AI Chatbots for WhatsApp & Telegram

- WhatsApp Business API – Automate customer queries, send updates, and handle orders.

- Twilio Autopilot – AI chatbot for WhatsApp, SMS & voice interactions.

- BotPress – Open-source chatbot platform with AI-powered automations.

3. AI Automation Tools for Email & CRM

- HubSpot & Zoho CRM – AI-powered chat + CRM for customer relationship management.

- ActiveCampaign – AI email automation + chatbot for follow-ups & engagement.

- Mailchimp & ConvertKit – AI-driven email sequences & customer nurturing.

5. How to Implement AI Chatbots in Your Business?

Step 1: Define Your Chatbot's Purpose

- Answer FAQs

- Guide users to resources

- Schedule calls/meetings

- Recommend products/courses

- Automate sales & support

Step 2: Choose a Platform & Build Your Chatbot

- Pick the right AI chatbot tool based on your platform (website, WhatsApp, social media).

Step 3: Train the AI with Relevant Data

- Use FAQs, past customer interactions, and real-world queries to train your bot.

Step 4: Integrate AI Chatbot Across Multiple Channels

- Ensure seamless experience on websites, social media, WhatsApp, and email.

Step 5: Test & Optimize for Better Performance

- Track chatbot performance and continuously improve responses.

6. AI Chatbots in Action: Use Cases for Engineers

Case 1: Engineering Course Sales

- Problem: Engineers struggle to manage leads and answer repetitive questions.

- Solution: AI chatbot answers FAQs, collects emails, and sends personalized course recommendations.

- Result: 40% increase in course enrollments.

Case 2: AI-Driven Support for Engineering Services

- Problem: Customers need 24/7 support for software/hardware queries.

- Solution: AI chatbot on WhatsApp provides instant troubleshooting & solutions.

- Result: 60% reduction in support tickets.

Case 3: Upselling Engineering Consultancy Services

- Problem: Clients drop off after initial engagement.

- Solution: AI chatbot follows up with tailored service recommendations.

- Result: 25% increase in repeat business.

7. Future Trends in AI-Powered Customer Service

- AI Chatbots with Emotional Intelligence – Detects emotions & responds empathetically.

- AI Voice Assistants for Engineers – Hands-free AI-powered solutions.

- Hyper-Personalized AI Recommendations – Tailors responses based on user behavior.

- AI-Powered Video Chatbots – Interactive AI-powered video support.

8. Action Plan: Implement AI Chatbots Today!

- Step 1: Choose a chatbot platform (WhatsApp, website, email, social media).

- Step 2: Define its key functions (FAQ, lead generation, sales, support).

- Step 3: Train the AI chatbot with real customer queries.

- Step 4: Automate customer interactions across multiple channels.

- Step 5: Monitor & optimize chatbot performance.

Ready to automate customer service with AI?

8.3. Digital Resources: PDFs, Videos, Audio, and AI-Generated Content

Leveraging AI-Powered Digital Content for Engineers

1. Why Use Digital Resources?

- Enhances Learning – People absorb information differently (reading, watching, listening).

- Increases Engagement – Visuals, animations, and interactive content boost understanding.

- Scalability – Once created, digital resources can be shared repeatedly without extra effort.

- AI Optimization – AI tools can generate, automate, and personalize content efficiently.

- Monetization – Digital products can be sold, licensed, or used for lead generation.

2. Types of Digital Resources & How to Create Them

PDFs – Structured Learning & Documentation

Use Cases:

- eBooks & Guides (e.g., "Mastering AI in Engineering")

- Whitepapers & Reports (Industry insights, case studies)

- Checklists & Worksheets (Action plans, templates)

- Course Notes & Study Materials

Tools to Create PDFs:

- Canva – Design beautiful PDFs with templates.

- Adobe Acrobat – Convert, edit, and enhance PDFs.

- Notion & Evernote – Create structured documents.

- AI Tools: ChatGPT & Jasper AI – Generate content for PDFs.

AI Advantage:

- AI can summarize long texts, convert voice notes into text, and even generate reports.

Videos – The Most Engaging Learning Medium

Use Cases:

- Explainer Videos – Break down complex topics with animations.

- Tutorials & Demos – Show how to apply engineering concepts.

- Webinars & Live Sessions – Interactive teaching for engineers.

- Promotional Videos – Market your courses or services.

Tools to Create Videos:

- Camtasia & ScreenPal – Screen recording & editing.

- CapCut & InShot – Easy video editing for social media.

- Descript – AI-powered video editing & transcription.

- Synthesia AI – Create AI-generated videos with virtual avatars.

AI Advantage:

- AI can generate video scripts, auto-edit clips, add subtitles, and create voiceovers.

Audio – Learning On the Go

Use Cases:

- Podcasts – Share industry insights & interviews.

- Audiobooks – Convert written content into audio format.

- Voice Notes & Q&A – Answer engineering queries with voice messages.

- AI-Generated Voiceovers – Convert text into professional audio.

Tools to Create Audio Content:

- Audacity & GarageBand – Record & edit audio.

- Descript – AI-powered transcription & voice cloning.

- ElevenLabs & Murf AI – AI-generated voiceovers.

- Anchor by Spotify – Create & distribute podcasts.

AI Advantage:

- AI can transcribe, generate natural-sounding voices, and even clone your voice for scalability.

AI-Generated Content – The Future of Digital Resources

AI can automate, optimize, and personalize content creation, saving time while maintaining quality.

AI Tools for Content Generation:

- ChatGPT & Jasper AI – Generate articles, reports, and learning materials.

- Pictory AI & Synthesia – Convert text into AI-powered videos.

- ElevenLabs & Murf AI – AI voiceovers and podcasts.

- Canva Magic Write & MidJourney – AI-generated graphics & presentations.

How AI Can Help Engineers?

- Auto-generate reports from raw data.

- Create personalized learning paths for students.

- Enhance engagement with interactive AI-based tutorials.

3. How to Monetize Digital Resources?

- Sell Digital Products – eBooks, courses, templates, and video lessons.

- Subscription-Based Access – Premium content membership sites.

- Affiliate Marketing – Promote relevant tools and earn commissions.

- Sponsorships & Ads – Earn through YouTube, podcasts, and blogs.

4. Action Plan: Create & Deploy Digital Content

- Step 1: Choose the format (PDF, video, audio, AI content).

- Step 2: Use AI tools to create and enhance your content.

- Step 3: Distribute via your website, social media, and LMS.

- Step 4: Monetize through sales, ads, or memberships.

- Step 5: Automate and scale using AI.

Engineers, it's time to leverage AI-powered digital content!

8.4. Building Long-Term Relationships & Community Growth

Why Engineers Must Focus on Relationships & Community

1. Why Long-Term Relationships Matter?

A loyal network of clients, peers, and industry experts brings:

- Trust & Credibility – People work with those they trust.

- Repeat Business – Clients return when they value your expertise.

- Word-of-Mouth Referrals – Happy clients become your best marketers.

- Collaboration Opportunities – Growth happens faster when you connect with like-minded professionals.

- Community Influence – A strong presence positions you as a thought leader.

If you only focus on short-term transactions, you'll constantly chase new opportunities instead of attracting them.

2. Strategies to Build Long-Term Relationships

Engage with Your Audience Regularly

- Be visible where your audience is—LinkedIn, industry forums, webinars.

- Respond to comments, messages, and emails personally.

- Share insights, answer questions, and provide genuine value.

Build Authentic Connections

- Be a giver, not just a taker – Offer value before expecting something in return.

- Celebrate milestones – Recognize achievements of clients and peers.

- Follow-up consistently – A simple check-in message strengthens relationships.

Provide Exceptional Service & Support

- Overdeliver – Always give more value than expected.

- Be accessible – Quick responses build trust.

- Create a knowledge hub – FAQs, video tutorials, and AI-powered support improve customer experience.

Thought Leadership & Community Engagement

- Write Blogs & Articles – Share knowledge to establish expertise.

- Host Webinars & Q&A Sessions – Engage your audience directly.

- Create a Newsletter – Stay connected with updates, tips, and insights.

- Public Speaking & Podcasts – Strengthen credibility in your niche.

3. Community Growth: How to Build a Thriving Network?

Create a Dedicated Community for Engineers

Engineers learn best from other engineers. Create a LinkedIn group, Discord server, Telegram channel, or forum to encourage discussions and knowledge sharing.

Why?

- A dedicated space makes it easier to interact with your audience.

- Members can learn from each other and share opportunities.

- Builds a support network for troubleshooting, career growth, and industry updates.

Host Live Events & Meetups

- Virtual or in-person meetups strengthen relationships.

- Live Q&A sessions build trust and engagement.

- AI-powered webinars allow you to scale engagement efficiently.

Leverage AI for Community Growth

AI-powered tools help automate, personalize, and enhance community engagement:

- Chatbots & AI Assistants – Provide 24/7 support in communities.

- AI-Generated Content – Create tailored newsletters, discussion prompts, and insights.

- Automated Engagement – Schedule posts, reminders, and event invites.

4. Monetizing Community Growth

Once you've built a strong, engaged network, monetization happens organically:

- Premium Memberships – Offer exclusive content & benefits.

- Consulting & Coaching – Provide personalized guidance.

- Affiliate & Sponsored Content – Partner with brands that add value to your audience.

- Online Courses & Webinars – Sell specialized knowledge.

5. Action Plan: Build & Sustain Relationships for Growth

- Step 1: Identify your core audience & where they engage the most.

- Step 2: Start genuine conversations, offer value first.

- Step 3: Create a community space (LinkedIn group, Telegram, Discord, etc.).

- Step 4: Use AI to automate engagement and personalize interactions.

- Step 5: Host live events, share insights, and collaborate for growth.

- Step 6: Monetize organically by offering premium value.

Start today!

********** MONETIZE **********

CHAPTER 9

WHY ENGINEERS MUST MASTER MONETIZATION

9.1 The shift from job-based income to expertise-based income

The shift from job-based income to expertise-based income is one of the biggest transformations happening in today's economy.

Engineers and professionals who recognize and adapt to this change will not only future-proof their careers but also unlock financial freedom, personal fulfillment, and meaningful impact.

Understanding the Shift

1. **Job-Based Income (Traditional Model)**

   o   Earned by exchanging time and skills for a salary.

   o   Limited by employer, industry, or market conditions.

   o   Growth is dependent on promotions, increments, or job changes.

   o   High risk of obsolescence due to automation and AI.

2. **Expertise-Based Income (New Model)**

- o Earned by monetizing specialized knowledge, skills, and experiences.

- o Not limited to a single employer—can be leveraged across multiple platforms.

- o Growth is exponential through content creation, consulting, training, and digital products.

- o Highly adaptable to industry shifts and technological advancements.

Why the Shift is Happening

- Automation & AI: Many traditional job roles are being replaced or reshaped by AI. Expertise, especially in niche areas, remains valuable.

- Internet & Digital Platforms: Knowledge can now be shared globally through YouTube, LinkedIn, online courses, and coaching programs.

- Entrepreneurial Mindset: More professionals are realizing they can monetize their skills beyond a job.

- Work-Life Freedom: Expertise-based income allows individuals to work on their terms rather than being tied to a 9-to-5 job.

How to Transition from Job-Based to Expertise-Based Income

1. **Identify Your Niche Expertise:**

   - What problems can you solve?

   - What unique knowledge or experience do you have?

   - What do people seek advice from you about?

2. **Build Your Authority & Brand:**

   - Create content on LinkedIn, YouTube, Medium, or a personal blog.

   - Write a book, start a podcast, or host webinars.

   - Share success stories and case studies.

3. **Monetize Your Expertise:**

   - Offer consulting services.

   - Create and sell online courses.

   - Conduct paid workshops and masterclasses.

   - Build membership communities or subscription models.

4. **Leverage Technology & Systems:**

   - Use AI tools to automate content creation and marketing.

   - Build a personal website and email list.

- o Set up digital distribution for courses, books, and coaching programs.

5. **Expand & Scale:**

   - o Collaborate with other experts.

   - o License your frameworks and methodologies.

   - o Develop multiple income streams from your expertise.

The future belongs to those who can package and present their expertise effectively. Engineers and professionals who develop their unique voice, share valuable insights, and monetize their knowledge will gain financial independence and purpose-driven success.

9.2. Overcoming the Mindset Barriers to Earning More

Engineers are great at solving technical problems, but when it comes to increasing income, many get stuck—not because of lack of skills, but because of limiting beliefs. If you want to break free from financial stagnation, you first need to upgrade your mindset.

Common Mindset Barriers That Hold Engineers Back

1. "My salary depends on my company, not me."

Truth: Your earning potential is in your hands. The more value you create, the more you can demand—whether in a job, consulting, or business.

2. "I don't have time to focus on earning more."

Truth: You don't need extra time; you need better priorities. Automate repetitive tasks, delegate, and invest time in high-value activities (building expertise, networking, or creating content).

3. "I need another degree or certification before I can charge more."

Truth: Companies and clients don't pay for certificates—they pay for results. Focus on applying what you already know and demonstrating impact.

4. "I'm not good at sales or marketing."

Truth: Sales is not about manipulation—it's about solving problems.

If you can explain how your expertise helps people or businesses, you're already selling.

5. "I don't deserve to earn more."

Truth: If your skills solve valuable problems, you deserve to be compensated well. Money is a byproduct of the impact you create.

How to Break These Barriers and Unlock Higher Earnings

Shift from Employee Mindset to Value-Creation Mindset

Stop thinking like someone who just executes tasks. Start thinking like someone who delivers solutions that save time, reduce costs, or drive innovation.

Monetize Your Expertise Beyond a Job

Your salary isn't the only way to earn. Start a consulting gig, write a book, launch a course, or coach aspiring engineers. Create multiple income streams.

Invest in Skills That Directly Impact Income

Not all skills are equal. Prioritize skills that boost your earning potential:

- Communication & Influence (to negotiate better pay)

- AI & Automation (to stay relevant)

- Personal Branding (to attract better opportunities)

Surround Yourself with High Earners

Your income is influenced by the people you interact with. Connect with those who have broken income barriers—learn from their mindset, strategies, and habits.

Your income is a reflection of the problems you solve and the mindset you hold. If you overcome these barriers, you won't just earn more—you'll unlock financial freedom and professional fulfillment.

Engineers, are you ready to break through and claim the income you deserve?

9.3. The Power of Financial Independence for Engineers

As engineers, we spend years mastering technology, problem-solving, and optimizing systems. But when it comes to our own financial growth, many of us still operate in survival mode, relying solely on a paycheck.

Financial independence is not just about money—it's about freedom, choices, and control over your career and life. Let's explore why it's essential and how you can achieve it.

Why Financial Independence Matters for Engineers

Freedom from Job Dependency

Relying only on a salary makes you vulnerable to layoffs, automation, and market fluctuations. When you achieve financial independence, you're no longer at the mercy of employers—you work because you want to, not because you have to.

More Innovation, Less Stress

When financial worries disappear, you can focus on high-impact work, take creative risks, and even start your own ventures without fear of failure.

Ability to Choose Projects You Love

With financial security, you can work on what excites you, whether it's deep tech, sustainability, or futuristic innovations, instead of just chasing salaries.

Stronger Negotiation Power

Financial independence gives you the confidence to demand your true worth, switch jobs strategically, or even walk away from toxic work environments.

Opportunities to Give Back & Mentor

When you're not worried about money, you can contribute to society, mentor young engineers, and create impact-driven initiatives.

How Engineers Can Achieve Financial Independence

1. Build Multiple Income Streams

- Relying on one paycheck is risky. Create other income sources like consulting, online courses, investing, or even an engineering side business.

2. Master High-Income Skills

- Beyond technical expertise, develop communication, leadership, and business skills—these directly impact earning potential.

3. Invest & Grow Your Money

- Learn about stocks, real estate, and passive income strategies. Make your money work for you rather than just working for money.

4. Monetize Your Knowledge

- Your expertise is valuable. Engineers can earn beyond a job by teaching, writing, coaching, or creating industry-specific solutions.

5. Think Like an Owner, Not an Employee

- Instead of seeing yourself as just a worker, think like a creator—build systems, optimize earnings, and own your career path.

Engineer Your Freedom!

Financial independence is not a luxury—it's a necessity. The moment you take control of your income, investments, and knowledge monetization, you unlock true career freedom.

So, are you ready to stop surviving and start thriving?

CHAPTER 10

MAKING MONEY – TURNING EXPERTISE INTO INCOME

10.1. Finding High-Value Problems to Solve

Why Engineers Must Focus on Solving High-Value Problems

As an engineer, your ability to identify, analyze, and solve problems defines your career success. But not all problems are worth solving. Some are too small to create an impact, while others don't have market demand. The key to success lies in finding high-value problems—those that have a strong demand, significant impact, and potential for monetization.

When you solve a high-value problem, you:

- Create real impact – Your solutions drive change.

- Attract high-paying clients – Businesses and individuals pay for valuable solutions.

- Build credibility – Expertise grows when you solve meaningful challenges.

- Achieve financial growth – Solving big problems leads to big rewards.

1. What Makes a Problem "High-Value"?

A high-value problem has these key characteristics:

1. Urgency & Pain Point

- How painful is this problem for people?

- Are they actively looking for a solution?

- How much time, money, or energy do they lose because of this problem?

- Example: A slow manufacturing process that increases costs and delays is a high-value problem because companies actively seek solutions to improve efficiency.

2. Market Demand & Willingness to Pay

- Do businesses or individuals have the budget to solve this?

- Are competitors already solving it (indicating demand)?

- Do industry leaders discuss this challenge?

- Example: AI-powered automation is in high demand because businesses are willing to invest in technology that reduces manual work and increases productivity.

3. Scalability & Long-Term Impact

- Does solving this problem open doors to bigger opportunities?

- Can your solution be adapted for different industries?

- Is this problem growing due to technological or societal changes?

- Example: Cybersecurity in AI applications is a high-value problem because as AI adoption increases, security risks also rise—making it a long-term opportunity.

2. How to Identify High-Value Problems?

Observe Industry Trends & Pain Points

- Follow industry news, reports, and LinkedIn discussions to spot emerging challenges.

- Analyze what experts are discussing at conferences and webinars.

- Look at frequently asked questions in forums and communities (e.g., Reddit, Quora, GitHub).

Talk to Industry Experts & Clients

- Ask clients: "What is your biggest challenge right now?"

- Conduct surveys & interviews with engineers and decision-makers.

- Join networking events & mastermind groups to understand industry struggles.

Use Data & AI Tools for Market Research

- Use Google Trends & AI-driven analytics to see what topics are growing.

- Check job postings—companies reveal their problems through skill requirements.

- Analyze patents & research papers to predict future needs.

Reverse Engineer Competitor Solutions

- Study successful startups and businesses—what problems are they solving?

- Look at customer reviews and complaints to find gaps in existing solutions.

- Use tools like Crunchbase, CB Insights, and Product Hunt to discover emerging innovations.

3. Examples of High-Value Problems in Engineering

Mechanical & Manufacturing Engineering

- Reducing downtime in factories using predictive maintenance.

- Improving material efficiency to lower production costs.

- Automating design processes with AI-driven CAD tools.

Electrical & Electronics Engineering

- Optimizing battery performance for renewable energy storage.

- Enhancing chip design for AI-powered applications.

- Developing low-cost IoT solutions for smart cities.

Software & AI Engineering

- Automating repetitive coding tasks with AI.

- Building AI-powered cybersecurity tools.

- Enhancing user experience with adaptive AI-driven interfaces.

Civil & Structural Engineering

- Creating eco-friendly, cost-effective building materials.

- Implementing AI-based structural health monitoring.

- Developing smart traffic management systems for urban areas.

Biomedical & Healthcare Engineering

- AI-driven diagnostics for early disease detection.

- Wearable health monitoring devices.

- Personalized medicine through data-driven insights.

4. Turning a Problem into an Opportunity

Once you've identified a high-value problem, follow these steps:

- Validate It – Talk to potential users and confirm if they'd pay for a solution.

- Prototype & Test – Build a small-scale version and get feedback.

- Develop Your Expertise – Learn and refine your skills to solve the problem effectively.

- Create & Deploy – Turn your solution into a product, service, or business.

- Communicate & Monetize – Market your solution through strategic selling.

5. Action Plan: Your Next Steps

- Step 1: Research industry trends and identify 3-5 common pain points.

- Step 2: Talk to experts and validate if these problems have market demand.

- Step 3: Choose one high-value problem to focus on.

- Step 4: Start developing a solution strategy based on your expertise.

- Step 5: Take action—prototyping, testing, and launching!

What high-value problem are you solving today? Let's take action

10.2. Creating & Delivering Solutions

As an engineer, your true value lies in your ability to create solutions that solve real-world problems. But just having an idea is not enough—you must design, develop, and deliver solutions in a way that is practical, scalable, and profitable.

In this section, we'll break down how you can systematically create high-impact solutions and deliver them effectively to your audience.

1. The Process of Creating a Solution

Step 1: Define the Problem Clearly

Before jumping into development, ensure you understand the problem in depth. Use the 5W1H Framework:

- What is the exact problem?

- Why is it important to solve?

- Who is affected by it?

- Where does this problem occur?

- When does it become critical?

- How is the problem currently being addressed (if at all)?

Example: If you are working on an AI-powered defect detection system for manufacturing, ask:

- What defects cause the most losses?

- Why do existing methods fail?

- Who will use this solution (factory operators, quality engineers, etc.)?

Step 2: Brainstorm & Validate Possible Solutions

Now, think of different ways the problem can be solved. Consider:

- Feasibility – Can it be built within available resources?

- Market Demand – Will people pay for it?

- Innovation Potential – Does it improve upon existing methods?

Tip: Use customer interviews, competitor analysis, and AI-driven market research to validate your idea.

Step 3: Prototype & Test

Once you have a concept, build a prototype or minimum viable product (MVP).

- For physical products: Use 3D modeling, CAD, or rapid prototyping tools.

- For software solutions: Develop a beta version and test with early users.

- For consulting or service-based solutions: Conduct pilot sessions with a small group.

Example: If you're developing an AI-based traffic management system, create a simulation using real-world data before full-scale deployment.

Step 4: Get Feedback & Improve

Once you test the prototype, collect user feedback:

- What works well?

- What needs improvement?

- Are there any unexpected issues?

Use iterative development—continuously refine and upgrade the solution based on insights.

2. Delivering Your Solution Effectively

Once your solution is ready, how you deliver it determines its success. Delivery isn't just about handing over a product; it's about ensuring your audience understands, trusts, and benefits from your solution.

A. Choosing the Right Delivery Model

Your delivery approach depends on the type of solution:

- Physical Products – Manufacturing, supply chain, distribution networks.

- Software Solutions – SaaS (Software as a Service), downloads, licensing.

- Digital Products – Courses, e-books, templates, AI-driven tools.

- Consulting/Coaching – Webinars, workshops, one-on-one sessions.

- Hybrid Models – Combining digital tools with hands-on services.

Example: If you're offering an AI-powered engineering design tool, you can deliver it as:

- A subscription-based SaaS product (monthly access).

- A one-time purchase software with future upgrades.

- A customized consulting service where you help companies integrate AI into their workflows.

B. Communicating Your Solution's Value

To ensure your solution reaches the right audience, you need clear messaging:

- What problem does it solve?

- How is it better than existing solutions?

- What measurable benefits does it provide?

- Why should someone invest in it NOW?

Example Messaging for an AI-Based Predictive Maintenance System:

"Our AI system monitors machines." (Too generic)

"Reduce machine downtime by 40% and save $50,000 annually with AI-powered predictive maintenance." (Clear, impact-driven)

Tip: Use case studies, testimonials, and demo videos to build credibility.

C. Scaling & Optimizing Delivery

Once your solution gains traction, focus on efficiency and scalability:

- Automation: Use AI tools for marketing, customer support, and delivery.

- Partnerships: Collaborate with industry players to expand reach.

- Community Building: Engage with users via forums, webinars, and online groups.

- Continuous Improvement: Regular updates based on market trends and feedback.

3. Action Plan: Your Next Steps

- Step 1: Identify a high-value problem in your domain.

- Step 2: Brainstorm possible solutions and validate with potential users.

- Step 3: Develop a prototype or MVP and gather feedback.

- Step 4: Choose the best delivery model for your solution.

- Step 5: Communicate value clearly and optimize for scalability.

What solution are you working on? Let's turn it into a success!

10.3. Business Models for Engineers in the AI Era

In the AI-driven world, engineers are no longer limited to traditional jobs. Instead, they have multiple pathways to monetize their expertise, build scalable businesses, and create long-term value. The key is to align your engineering skills with market demand while leveraging AI for efficiency and growth.

Let's explore the best business models for engineers in today's AI-powered economy.

1. AI-Enabled Consulting & Services

What It Is

As an engineer, you can offer specialized consulting services to businesses looking to integrate AI and automation. Many companies struggle with AI adoption due to a lack of expertise—this is where you step in.

How to Monetize

- One-on-One Consulting: Help businesses implement AI solutions in their industry.

- Project-Based Services: Offer AI model development, automation solutions, or technical advisory.

- Retainer Model: Provide ongoing AI consulting services for monthly fees.

Example: An AI engineer specializing in machine learning can offer consulting to manufacturing firms for predictive maintenance solutions.

2. AI-Powered SaaS (Software as a Service)

What It Is

Build a subscription-based software platform that solves a specific problem using AI. This model provides recurring revenue and scalability.

How to Monetize

- Subscription Plans: Charge users a monthly/yearly fee.

- Freemium Model: Offer free basic features with premium upgrades.

- Enterprise Licensing: Provide AI tools to large companies for bulk licensing.

Example: An AI engineer develops a resume screening software for HR teams, using AI to analyze job applications.

3. Online Courses & Digital Products

What It Is

Engineers with expertise in AI, automation, or any other domain can create educational content and sell it online.

How to Monetize

- Self-Paced Courses: Sell recorded courses on platforms like Udemy, Teachable, or Kajabi.

- Live Training: Offer premium live webinars or coaching.

- E-Books & Technical Guides: Sell digital resources with in-depth knowledge.

Example: A robotics engineer creates an online course on AI in industrial automation and sells it on Coursera.

4. AI-Powered E-Commerce & Affiliate Marketing

What It Is

Leverage AI-driven tools to run a profitable e-commerce business or affiliate marketing model without handling inventory.

How to Monetize

- Dropshipping: Sell AI-recommended products without storing inventory.

- Affiliate Marketing: Promote AI-based tools/software and earn commissions.

- AI-Driven Print-on-Demand: Use AI tools for automated product design.

Example: An engineer with an interest in AI hardware creates an e-commerce store selling AI-powered IoT gadgets.

5. AI-Driven Freelancing & Microservices

What It Is

Engineers can provide on-demand AI and engineering services on freelancing platforms.

How to Monetize

- Hourly Rates: Charge clients based on work hours.

- Project-Based Pricing: Charge a flat fee for AI-related projects.

- Gig Economy Work: Offer specialized microservices in AI, coding, automation, etc.

Example: A data scientist offers AI model development services on Upwork and Fiverr.

6. AI-Based Product Development & Patents

What It Is

Engineers can create and patent AI-driven innovations, then license or sell them.

How to Monetize

- Patent & License: Develop AI-powered hardware/software and license it to businesses.

- Sell Intellectual Property: Sell your AI-based inventions to companies.

- Startup Route: Build a tech startup based on your innovation.

Example: An AI engineer develops a new algorithm for medical image analysis and licenses it to healthcare startups.

7. Hybrid Model: Blending Multiple Streams

Instead of choosing just one model, engineers can combine multiple business approaches for higher earnings.

Example Hybrid Approach:

- Step 1: Start as a consultant to validate your expertise.

- Step 2: Create AI-based SaaS or digital products based on client needs.

- Step 3: Build an online course and generate passive income.

- Step 4: Scale the business through AI automation and community growth.

Example: A machine learning engineer provides AI consulting, launches an automated SaaS tool for AI model deployment, and sells AI training courses online.

Action Plan: How to Get Started Today

1. Identify your expertise area. Which AI or engineering field excites you?

2. Choose a business model that suits your skills and market demand.

3. Develop a Minimum Viable Product (MVP). Test your solution before scaling.

4. Leverage AI tools for automation, marketing, and efficiency.

5. Build a strong online presence to attract clients and customers.

Which business model excites you the most? Let's build your AI-driven success story!

CHAPTER 11

MANAGING MONEY – SMART FINANCIAL STRATEGIES

11.1. Categorizing Income for Financial Stability

As an engineer aiming for financial independence, it's essential to manage money strategically. Financial stability doesn't come from just making money—it comes from organizing, allocating, and managing your income effectively. Without a clear system, you might overspend, under-save, or fail to scale your financial growth.

To achieve this, you must categorize your income into different financial buckets, ensuring that every rupee (or dollar) has a purpose. Let's dive into the structured approach that will help you grow and sustain wealth.

Why Categorizing Income is Crucial

1. Prevents Overspending – Allocating money to different categories helps control unnecessary expenses.

2. Ensures Financial Security – Setting aside money for emergencies and investments safeguards your future.

3. Helps in Wealth Creation – Dividing income strategically enables smart investing and long-term growth.

4. Builds Financial Discipline – When money is pre-assigned to categories, you develop a habit of responsible spending.

5. Prepares for Business Growth – Engineers who aspire to launch businesses or side projects can allocate funds wisely.

The 6-Account Money Management System

A simple way to categorize your income is by using a 6-account system. Every time you earn money, divide it into these accounts to ensure financial balance.

1. Essentials Account (50%) – "Survival & Living"

Purpose: Covers daily living expenses like rent, food, transportation, and utilities.

Ideal Allocation: 50% of your income

Includes:

- Rent/Mortgage

- Groceries & Essentials

- Transport & Fuel

- Utility Bills

- Insurance

Tip: If your essential expenses exceed 50%, find ways to cut unnecessary spending or increase your income.

2. Financial Freedom Account (10%) – "Wealth-Building & Investments"

Purpose: This is your "never touch" account for investments that generate passive income.

Ideal Allocation: 10% of your income

Includes:

- Stocks & Mutual Funds

- Real Estate Investments

- AI-Based Investment Platforms

- Startup or Side Business Funding

Tip: The money here should work for you—focus on assets that generate ongoing income.

3. Savings & Emergency Fund (10%) – "Security & Peace of Mind"

Purpose: This account is your safety net for unexpected expenses.

Ideal Allocation: 10% of your income

Includes:

- Emergency Fund (6-12 months of expenses)

- Medical Expenses

- Sudden Job Loss Protection

Tip: Always keep this money liquid—it should be accessible when needed.

4. Education & Skill Growth (10%) – "Upgrading Your Knowledge"

Purpose: Continuous learning is essential for engineers to stay ahead in the AI era.

Ideal Allocation: 10% of your income

Includes:

- Online Courses (AI, Automation, Business)

- Books & Learning Materials

- Certifications & Workshops

- Coaching & Mentorship

Tip: The more you invest in knowledge, the more valuable you become in the market.

5. Fun & Recreation (10%) – "Enjoying Life Without Guilt"

Purpose: Money is meant to be enjoyed! This account allows guilt-free spending on things you love.

Ideal Allocation: 10% of your income

Includes:

- Travel & Vacations

- Entertainment & Hobbies

- Dining Out & Leisure Activities

Tip: If you don't spend this money, don't save it here—use it to enjoy life!

6. Contribution & Charity (10%) – "Giving Back to Society"

Purpose: Giving creates abundance and purpose in life.

Ideal Allocation: 10% of your income

Includes:

- Supporting NGOs & Charities

- Helping Family or Friends in Need

- Community Projects

Tip: When you give with an open heart, you attract even more abundance in return.

Practical Implementation: How to Start

Step 1: Open separate bank accounts or use financial apps to allocate funds automatically.

Step 2: Every time you earn, split your income into these categories.

Step 3: Track and review your spending monthly to ensure financial discipline.

Step 4: Adjust the percentages if needed, but always prioritize investment and security.

Engineers' Money Mindset: Think Long-Term

Money loves clarity and structure. If you manage it well, you will never feel financially trapped.

Aim for financial independence. Categorizing income is the first step towards freedom from financial stress.

Start today. Even if you can't allocate exactly 10% to each category, begin with what you can and scale up.

Take Action Now!

What's the first step you will take to categorize your income?

Are you already practicing any of these money management strategies?

Let's build financial stability and growth together!

11.2. AI-Based Financial Management Tools for Engineers

Managing money wisely is just as important as making it. With AI-driven financial tools, you can track, analyze, and optimize your finances effortlessly.

These tools leverage automation, machine learning, and predictive analytics to help engineers budget, invest, and grow wealth strategically.

Why Use AI for Financial Management?

- Automates Expense Tracking – No more manual logging; AI categorizes your spending in real time.

- Optimizes Savings & Investments – AI recommends smart investment strategies based on your financial goals.

- Provides Predictive Insights – Forecasts income, expenses, and potential financial risks.

- Eliminates Human Errors – Ensures precision in tax calculations, budgeting, and financial planning.

- Saves Time & Effort – Engineers can focus on growth while AI handles financial complexities.

Best AI-Based Financial Tools for Engineers

1. AI-Powered Budgeting & Expense Tracking

Tools: Mint, YNAB (You Need A Budget), PocketGuard

How It Helps:

- Tracks income and expenses automatically.

- Provides personalized budgeting suggestions.

- Alerts for overspending and saving opportunities.

Tip: Set AI-driven spending limits to control impulsive purchases.

2. AI for Investment & Wealth Growth

Tools: Wealthfront, Betterment, SigFig, Zerodha Streak (for India)

How It Helps:

- Robo-advisors suggest personalized investment strategies.

- AI monitors market trends and auto-adjusts portfolios.

- Helps engineers invest in stocks, mutual funds, real estate, and crypto.

Tip: AI-based index funds and ETFs are great for low-risk, automated investing.

3. AI for Tax & Financial Planning

Tools: TurboTax, ClearTax, H&R Block, QuickBooks AI

How It Helps:

- Automates tax filing with error-free calculations.

- Identifies deductions and tax-saving opportunities.

- Generates financial reports for better planning.

Tip: AI tax tools save time and prevent penalties due to errors.

4. AI-Powered Personal Finance Assistants

Tools: Cleo, Olivia, Digit, Plum

How It Helps:

- AI bots analyze spending habits and suggest savings goals.

- Provides financial coaching and spending tips.

- Helps you automate savings without even thinking about it.

Tip: Use AI assistants to automate weekly savings deposits.

5. AI for Business Finance & Accounting

Tools: Xero, FreshBooks, Zoho Books, QuickBooks AI

How It Helps:

- Manages business finances, invoicing, and cash flow.

- AI tracks project expenses and generates reports.

- Automates bookkeeping for freelancers & business owners.

Tip: Engineers running consulting businesses should integrate AI-based accounting tools.

How Engineers Can Leverage AI for Financial Success

- Automate Your Budget – Use AI tools to track and manage spending.

- Invest Smartly – Let robo-advisors handle stock investments and asset allocation.

- Plan Taxes in Advance – AI tax software ensures compliance and deductions.

- Optimize Business Finances – AI tools simplify accounting for freelancers and entrepreneurs.

- Set Financial Goals – AI helps achieve milestones like buying a home, funding education, or retiring early.

Take Action Now!

- Which AI financial tool are you excited to try first?

- Are you already using any AI-powered tools for managing your money?

Start automating your finances today and build lasting wealth!

11.3. Money Management Habits & Building a Wealth Mindset for Engineers

As an engineer, you are trained to solve problems with logic and precision, but have you applied the same approach to managing your money? Financial success is not just about earning more; it's about managing, growing, and sustaining wealth. This requires the right habits and a strong wealth mindset.

Why Engineers Struggle with Money Management

Many engineers earn well but struggle to save or grow their money due to:

- Lack of financial literacy and planning.

- Spending without budgeting or tracking.

- Not investing early and missing out on compounding.

- Fear of taking financial risks.

- No clear strategy to multiply income streams.

The good news? With the right habits and mindset, you can turn your income into lasting wealth!

Money Management Habits for Engineers

1. Track Every Rupee/Dollar You Spend

- Use AI-powered finance apps like Mint, YNAB, PocketGuard to automate tracking.

- Categorize expenses: Needs (50%) – Wants (30%) – Savings/Investments (20%).

- Set a monthly spending limit and review it weekly.

Pro Tip: If you don't track your money, you'll never know where it's going!

2. Follow the 6-Account Money Management System

Divide your income into different accounts to ensure financial stability:

1. Necessities (50%) – Rent, food, bills, transport.

2. Financial Freedom (10%) – Investments for passive income.

3. Long-Term Savings (10%) – Emergency funds, big purchases.

4. Education & Growth (10%) – Courses, books, mentors.

5. Fun & Recreation (10%) – Travel, entertainment.

6. Giving (10%) – Charity, donations, contributions.

Pro Tip: Automate transfers to these accounts for discipline.

3. Invest Early & Consistently

- Start investing a fixed percentage of your income every month.

- Long-term focus: Stocks, Index Funds, Mutual Funds, Real Estate, and AI-driven investments.

- Use AI tools like Zerodha Streak, Wealthfront, Betterment to automate investing.

Pro Tip: The earlier you invest, the more you benefit from compounding interest.

4. Have an Emergency Fund

- Save at least 6-12 months of living expenses in a separate account.

- Use high-yield savings accounts or fixed deposits for easy access.

Pro Tip: This protects you from unexpected job loss, medical issues, or financial crises.

5. Create Multiple Income Streams

- Side Hustles: Freelancing, consulting, online courses, YouTube, blogging.

- Passive Income: Stocks, dividends, real estate, affiliate marketing.

- Monetize Your Expertise: Sell digital products, e-books, or AI-powered services.

Pro Tip: Never depend on a single source of income. Build financial security through multiple streams.

Developing a Wealth Mindset for Engineers

Mindset is everything! You can make money, manage it well, and grow it exponentially if you think like a wealth creator.

1. Shift from Employee to Investor & Creator Mindset

- Old Mindset: Earn & spend money on liabilities.

- New Mindset: Invest in assets that generate income.

Example: Instead of just earning from a job, create products, courses, or automated services that bring money continuously.

2. Focus on Value Creation, Not Just Earning

Wealthy engineers don't chase salary hikes; they create valuable solutions.

- Find high-value problems to solve in your industry.

- Develop products, patents, and innovations that people pay for.

Pro Tip: The more problems you solve, the more wealth you create!

3. Overcome the Fear of Investing

Many engineers hesitate to invest due to fear of risk.

- Start small but start now.

- Educate yourself about stocks, crypto, AI investments, real estate.

- Use AI-driven tools for risk analysis & smart investing.

Pro Tip: Not investing is the biggest financial risk!

4. Surround Yourself with Financially Smart People

- Join wealth-building communities.

- Follow mentors & successful entrepreneurs.

- Learn from experts in finance, business, and investments.

Pro Tip: Your financial success depends on who you learn from!

Take Action & Redefine Your Financial Growth!

Which money management habit will you implement first?

What steps will you take to develop a wealth mindset?

Start managing money like an engineer and build unstoppable financial success!

CHAPTER 12

MULTIPLYING MONEY – SCALING & GROWTH

12.1. Exploring Multiple Income Streams for Engineers

As an engineer, you have high-income potential, but are you making the most of it? Relying on a single source of income—like a job—limits financial growth and security. To achieve true freedom and financial success, you need to diversify your income sources.

Let's explore multiple ways you can create, manage, and multiply your income without quitting your job immediately.

Why Multiple Income Streams Matter for Engineers

- Job security is not guaranteed – AI and automation are changing industries.

- Inflation is rising – A fixed salary won't keep up.

- More income = More freedom – Travel, better lifestyle, early retirement.

- Your skills are valuable – Monetize your expertise beyond your job.

If you are dependent on a single income source, you're one decision away from financial instability.

Solution? Create multiple streams of income!

7 Income Streams Every Engineer Should Have

1. Salary / Profit: Primary Job/Business Income

This is your main cash flow. Your goal should be to maximize it through expertise, career growth, and negotiations.

- Negotiate for better pay & promotions.

- Develop niche expertise to increase demand.

- Leverage AI & automation to boost productivity and performance.

Pro Tip: Never depend 100% on this income. Use it to build other income sources!

2. Freelancing & Consulting

If you have engineering skills, companies and individuals are willing to pay for your expertise.

- Offer services on Upwork, Fiverr, Toptal.

- Specialize in AI, software, electrical, mechanical, civil, or automation consulting.

- Provide solutions for design, simulation, coding, or troubleshooting.

Example: A mechanical engineer can offer CAD design services to global clients online.

3. Online Courses & Coaching

Engineers worldwide are looking for expertise and guidance. You can teach and earn by creating online courses.

- Sell courses on Udemy, Teachable, Kajabi, Thinkific.

- Conduct webinars & workshops on niche topics.

- Offer 1-on-1 coaching to junior engineers & students.

Example: A software engineer can teach Python for AI or Automation using IoT.

Pro Tip: AI tools like ChatGPT, Synthesia, and Descript can help create content faster.

4. Digital Products & E-books

If you can create knowledge-based content, you can make passive income.

- Write an engineering e-book or guide.

- Sell templates, Excel tools, AI automation scripts.

- Offer downloadable resources on Gumroad, Etsy, or Sellfy.

Example: An electrical engineer can sell PCB design templates & simulation models.

5. Affiliate Marketing & Partnerships

You can promote products, tools, and software related to your field and earn commissions.

- Join affiliate programs of Amazon, Coursera, LinkedIn Learning, Autodesk, MATLAB.

- Write blogs or make YouTube videos reviewing engineering software & tools.

- Recommend useful products on LinkedIn, Twitter, or personal blogs.

Example: A software engineer can earn commissions by promoting coding courses or AI tools.

6. Investments: Stocks, Crypto & Real Estate

Smart investments can make money work for you.

- Stock Market: Invest in tech, automation, AI-based companies.

- Crypto & Blockchain: Explore potential investments in Web3, AI-driven DeFi projects.

- Real Estate: Buy & rent out properties or invest in Real Estate Investment Trusts (REITs).

Pro Tip: Use AI-based investment platforms like Wealthfront, Zerodha, or Robinhood for smart investing.

7. E-commerce & AI-Driven Dropshipping

Selling products online is easier than ever. AI automates most processes!

- Dropshipping: Sell products without holding inventory via Shopify, WooCommerce.

- Print-on-Demand: Design engineering-themed T-shirts, mugs, posters on Redbubble, Teespring.

- Tech Gadgets & Tools: Sell niche engineering equipment or digital tools.

Example: A mechanical engineer can sell 3D-printed custom tools on Amazon.

Action Plan: Start Your First Side Income Today!

1. Pick ONE income stream that matches your skills & interests.

2. Dedicate 2-5 hours per week to start.

3. Use AI tools to automate and scale.

4. Track progress and refine your strategy.

5. Once stable, expand into other income streams.

Pro Tip: Focus on building passive income, so you earn while you sleep!

Which income stream will you start first?

12.2. Scaling Business/Job Opportunities Using AI

Engineers who master AI can scale their careers and businesses exponentially. Whether you are an entrepreneur, a freelancer, or an employee, AI-driven automation and data insights can help you increase efficiency, reduce costs, and expand opportunities like never before.

Let's explore how you can leverage AI to scale your career, business, or job.

Why Engineers Must Use AI for Growth?

- AI increases productivity – Automate repetitive tasks and focus on innovation.

- AI enhances decision-making – Data-driven insights help in smart planning.

- AI personalizes solutions – Customized services attract more clients.

- AI opens new career/business paths – Stay ahead of competitors.

Pro Tip: Engineers who use AI in their work will outperform those who don't!

5 Ways AI Can Scale Your Career & Business

1. AI for Automating Repetitive Engineering Tasks

Time is money! If you are doing the same tasks every day, AI can automate them so you can focus on higher-value work.

- Mechanical & Electrical Engineers: Use AI-driven CAD automation & simulations (e.g., SolidWorks AI, Ansys AI).

- Software Engineers: Automate coding with GitHub Copilot, ChatGPT, Tabnine.

- Civil Engineers: Use AI in design optimization, smart city planning.

Example: Instead of manually designing multiple prototypes, AI-driven simulation tools can generate optimized models instantly.

2. AI for Enhancing Decision-Making & Strategy

AI-powered data analysis tools can help you make smarter career or business decisions.

- Use ChatGPT, Jasper AI for market research.

- Predict trends using AI analytics tools like Tableau AI, IBM Watson.

- Improve engineering project planning with AI-based risk analysis tools.

Example: An entrepreneur selling engineering online courses can use AI to analyze trends and create high-demand courses.

3. AI for Scaling Client Outreach & Sales

AI-driven tools help engineers reach a larger audience and increase sales effortlessly.

- Automate email marketing using Mailchimp AI, ActiveCampaign.

- Use AI chatbots to handle customer queries 24/7 (e.g., Drift, ManyChat).

- Run AI-powered ads on Facebook, Google, and LinkedIn to attract high-quality clients.

Example: A freelancer offering AI automation consulting can use AI-powered LinkedIn outreach tools to connect with high-value clients worldwide.

4. AI for Personalized Learning & Career Growth

AI can help engineers upskill and grow faster.

- AI-powered learning platforms like Coursera AI, Udemy AI recommend personalized courses.

- Use ChatGPT, Claude AI as your 24/7 mentor for instant learning.

- AI-based skill assessment identifies gaps in expertise and suggests improvements.

Example: A mechanical engineer can learn AI-driven predictive maintenance through AI-curated courses.

5. AI for Creating & Scaling Digital Products

If you want to scale your business, AI makes product creation & delivery seamless.

- AI-generated content for blogs, courses, and videos (e.g., Jasper AI, Synthesia).

- AI-powered product design for eBooks, reports, and presentations.

- Automate product delivery through AI-driven LMS and customer service chatbots.

Example: A software engineer can use AI to create and sell coding templates, making passive income.

How to Get Started with AI for Scaling?

- Step 1: Identify areas where AI can save you time.

- Step 2: Choose AI tools that fit your industry.

- Step 3: Start small – Automate one task at a time.

- Step 4: Track performance and expand AI integration.

- Step 5: Scale your career/business with AI-driven insights.

Which AI tool will you start using today?

12.3. Cross-Selling, Up-Selling & Affiliate Marketing for Engineers

Engineers who master cross-selling, up-selling, and affiliate marketing can significantly boost their income by maximizing value for their clients. Whether you're selling a service, a digital product, or a physical product, these strategies help you generate more revenue from your existing customer base.

Let's break it down into simple, actionable steps.

1. Cross-Selling: Sell More by Offering Complementary Products

Cross-selling is about suggesting related products or services that enhance the primary purchase.

- For Freelancers & Consultants – If you offer a website development service, cross-sell SEO optimization or content writing.

- For Course Creators – If you sell a basic AI automation course, offer a bonus advanced AI course.

- For Product Creators – If you sell engineering design templates, cross-sell a customization service.

Example: A mechanical engineer selling 3D CAD templates can cross-sell custom design modifications or consulting sessions.

Action Step: Identify 3 additional services or products you can offer alongside your main product.

2. Up-Selling: Encourage Higher-Value Purchases

Up-selling is about offering a premium version or additional features of your product/service.

- For Freelancers & Consultants – If you offer a basic automation setup, upsell a full AI-driven business automation package.

- For Course Creators – If you sell a self-paced engineering course, upsell a live coaching program.

- For Product Creators – If you sell a basic engineering tool, upsell a bundle with exclusive features.

Example: A data engineer selling Python automation scripts can upsell a customized script development service.

Action Step: Create a tiered pricing strategy to offer higher-value options.

3. Affiliate Marketing: Earn Passive Income by Promoting Relevant Products

Affiliate marketing allows you to earn commissions by recommending products and services that you trust.

- Join Affiliate Programs – Sign up for Amazon Associates, Udemy, AI tools, software subscriptions.

- Create Content & Recommendations – Write blogs, make videos, or share recommendations via email.

- Leverage Your Expertise – Promote tools you personally use and provide value to your audience.

Example: A cloud engineer can promote AWS certifications or hosting services and earn commissions.

Action Step: Identify 3 products related to your niche that you can start promoting today.

4. How to Maximize Income with These Strategies?

- Analyze Customer Needs – Understand what they want beyond the first purchase.

- Bundle Smartly – Offer valuable cross-sells and up-sells that feel like a no-brainer.

- Leverage AI & Automation – Use AI-powered chatbots and email sequences to promote offers.

- Track & Optimize – Use analytics tools to measure what works best.

Which strategy will you apply first – Cross-Selling, Up-Selling, or Affiliate Marketing?

12.4. Leveraging AI for E-Commerce & Digital Product Expansion

In today's digital economy, AI is revolutionizing how engineers create, sell, and scale digital products and e-commerce businesses. Whether you're selling online courses, digital templates, software, or engineering tools, AI can help automate and optimize every step of the process.

Let's explore how you can leverage AI to expand your digital product business effectively.

1. AI-Powered Product Research & Trend Analysis

Before creating a product, you must identify market demand. AI tools help by analyzing trends, customer behavior, and competition.

- Use AI-driven tools like Google Trends, ChatGPT, and Helium 10 to analyze market gaps.

- Analyze competitor products to identify missing features or improvements.

- Use AI-based sentiment analysis to study customer reviews and refine your offerings.

Example: An electrical engineer wanting to sell Arduino project kits can use AI to find trending project ideas and optimize designs.

Action Step: Use AI-powered trend analysis tools to validate product demand.

2. AI-Driven Product Creation & Automation

AI can assist in creating digital products faster and more efficiently, reducing manual effort.

- For E-Books & Courses: Use AI like ChatGPT to generate content outlines, summaries, and interactive quizzes.

- For Software & Tools: Use AI-powered coding assistants like GitHub Copilot to accelerate software development.

- For Engineering Templates & Models: Use AI-driven design tools like AutoCAD AI and SolidWorks AI to generate optimized designs.

Example: A mechanical engineer can create an AI-powered design optimization tool for 3D printing enthusiasts.

Action Step: Identify one aspect of product creation that AI can automate.

3. AI-Optimized Digital Store & E-Commerce Setup

Your online store needs a smooth user experience, AI-powered recommendations, and smart automation.

- AI-Powered E-Commerce Platforms: Shopify AI, WooCommerce AI, and Magento AI help automate product recommendations and customer support.

- Chatbots & AI Assistants: Deploy AI chatbots like Drift, ChatGPT, or Tidio to handle FAQs and increase conversions.

- Personalized Shopping Experiences: Use AI-driven recommendation engines like Recombee or Dynamic Yield to offer personalized product suggestions.

Example: A software engineer selling custom coding scripts can use an AI-powered chatbot to guide buyers to the right product.

Action Step: Integrate an AI chatbot into your e-commerce site for automated support.

4. AI-Powered Marketing & Customer Engagement

AI makes marketing faster, smarter, and more effective by analyzing customer behavior and automating campaigns.

- AI Content Creation: Use AI tools like Jasper and Copy.ai to generate blog posts, product descriptions, and ad copies.

- AI Video Marketing: Use Synthesia or Pictory AI to create engaging video promotions without expensive production.

- AI-Powered Email Marketing: Automate personalized email sequences using tools like Mailchimp AI and HubSpot AI.

Example: A civil engineer selling structural analysis tutorials can use AI-generated YouTube videos to attract an audience.

Action Step: Use an AI content generation tool for your next blog, video, or email campaign.

5. AI-Based Pricing & Sales Optimization

AI-driven pricing tools analyze market demand, customer willingness to pay, and competitor pricing to help you set optimal prices.

- AI-Powered Dynamic Pricing: Use AI tools like Prisync or ProfitWell to adjust pricing based on demand and trends.

- AI-Based Upselling & Cross-Selling: Use AI-driven algorithms to recommend add-ons or higher-value products.

- AI Chatbots for Sales Assistance: Automate pre-sales questions and help customers choose the right product.

Example: A data engineer selling AI automation scripts can use AI-driven pricing models to offer discounts or bundle deals dynamically.

Action Step: Experiment with AI-powered pricing optimization for your digital products.

6. Scaling with AI-Driven Growth Strategies

Once your e-commerce or digital product business gains traction, AI can help scale effortlessly.

- AI-Powered Performance Analytics: Use AI tools like Google Analytics AI and Tableau AI to track conversion rates and optimize marketing.

- AI Chatbots for 24/7 Customer Support: Automate responses and reduce support workload.

- AI-Driven Localization & Translation: Expand globally by using AI-powered translation tools to reach international customers.

Example: An engineer selling AI-driven automation templates can use AI translation tools to sell worldwide.

Action Step: Identify one AI-driven scaling strategy to expand your reach.

The Future of AI in E-Commerce & Digital Product Expansion

AI is no longer an option—it's a necessity for engineers looking to scale their digital businesses. From product research to creation, marketing, and sales, AI can automate and optimize every step.

Which AI tool or strategy will you start implementing today?

12.5. Creating and Scaling Physical Products for Maximum Impact

In an era dominated by digital solutions, physical products still hold immense value—especially for engineers who want to create tangible solutions that solve real-world problems. Whether it's hardware, engineering tools, IoT devices, or innovative gadgets, AI and automation can help streamline product development, manufacturing, and scaling.

Let's explore how you can create, launch, and scale physical products effectively using AI-driven strategies.

1. Identifying Market Needs & Validating Product Ideas

Before creating a physical product, engineers must identify a problem that people are willing to pay for. AI-driven tools can help validate demand and analyze market trends.

- AI-Powered Market Research: Use AI tools like Google Trends, ChatGPT, and SEMrush to identify product demand.

- Customer Sentiment Analysis: Tools like Brandwatch AI analyze customer pain points from online reviews and forums.

- Rapid Prototyping & Feedback: Use AI-powered design tools to create digital prototypes and gather early feedback.

Example: A mechanical engineer designing a smart ergonomic desk can use AI to analyze common posture-related problems and optimize the design.

Action Step: Use AI market research tools to validate your physical product idea.

2. Designing & Prototyping with AI & Automation

Once the idea is validated, the next step is to design and prototype efficiently using AI-powered design tools.

- AI-Driven CAD Design: Tools like AutoCAD AI, SolidWorks AI, and Fusion 360 use machine learning to optimize product design.

- Generative Design: AI suggests optimized designs based on material strength, weight reduction, and cost efficiency.

- 3D Printing & Rapid Prototyping: AI-driven 3D printing software like Formlabs AI helps engineers iterate faster.

Example: An electronics engineer can use AI-driven PCB design tools like Altium AI to create optimized circuit boards.

Action Step: Try using AI-based CAD design tools to optimize your product's structure.

3. AI-Powered Manufacturing & Automation

To scale production, leveraging AI in manufacturing is key. AI helps optimize costs, reduce waste, and improve efficiency.

- AI-Driven Supply Chain Optimization: AI tools like Llamasoft AI help predict demand and optimize sourcing.

- Robotic Process Automation (RPA): AI-powered robots handle repetitive manufacturing tasks, increasing efficiency.

- Predictive Maintenance: AI-based sensors in factories prevent downtime by predicting machine failures.

Example: A robotics engineer manufacturing AI-powered drones can use AI-driven quality control systems for precision.

Action Step: Research AI-driven automation for manufacturing processes to improve efficiency.

4. Marketing & Selling Physical Products Using AI

Once the product is ready, AI-powered marketing and sales strategies can help engineers reach the right audience.

- AI-Powered E-Commerce: Platforms like Shopify AI and Amazon AI optimize listings, pricing, and recommendations.

- AI-Driven Ad Campaigns: Tools like Adzooma and Meta AI optimize advertising for better ROI.

- AI Chatbots for Sales Support: Automate customer inquiries with AI-powered chatbots like Drift and Intercom.

Example: A civil engineer selling modular home kits can use AI-driven ads to target eco-conscious customers.

Action Step: Use AI-powered marketing tools to launch and sell your physical product effectively.

5. Scaling & Expanding Your Physical Product Business

Once sales grow, scaling requires AI-powered automation and strategic expansion.

- AI-Optimized Logistics & Inventory: AI tools like Flexport AI manage supply chains and optimize shipping.

- AI-Based Customer Support: AI chatbots handle support tickets, reducing workload.

- Expanding to Global Markets: AI-driven translation tools like DeepL AI help localize products for international markets.

Example: A mechanical engineer selling AI-powered robotic arms can scale internationally by leveraging AI-driven logistics and customer support.

Action Step: Identify one AI-driven scaling strategy to expand your reach.

AI-Driven Growth for Physical Product Success

Engineers have an incredible opportunity to create and scale innovative physical products using AI. From idea validation and design to manufacturing, marketing, and scaling, AI-driven strategies can help optimize every step.

Which AI tool or strategy will you implement first?

****** BEYOND YOUR SUCCESS ******

FINAL CHAPTER

ENGINEERS AS THE ARCHITECTS OF A TECH-DRIVEN FUTURE

a) How Engineers Can Redefine Growth Beyond Financial Success

True growth is not just about financial success—it's about impact, innovation, and contribution. As engineers, we have the power to build a future that goes beyond profits and creates lasting value for society.

So, how can engineers redefine growth in a way that is holistic, meaningful, and fulfilling? Let's explore key areas where engineers can expand their impact.

1.Shift from Success to Significance

Many engineers focus on high salaries, promotions, and financial stability—which are important. But true growth happens when you shift your focus from personal success to societal significance.

While achieving career success is important, engineers should also strive for impact and legacy.

- Ask yourself:

- o How is my work improving lives?

- o Am I solving problems that truly matter?

- o What legacy do I want to leave as an engineer?

- Example: A software engineer building AI-driven automation can focus on solutions that improve accessibility for people with disabilities, rather than just creating high-paying software.

Action Step: Identify one way you can make your expertise more impactful beyond financial gain.

2.Solve Real-World Problems for a Better Future

Engineers have the mindset and skills to solve some of humanity's biggest challenges—including climate change, healthcare, education, sustainability, and infrastructure.

- Ways engineers can contribute:

- o Civil engineers can work on sustainable urban planning.

- o Electrical engineers can innovate in renewable energy solutions.

- o Mechanical engineers can design energy-efficient machinery.

- o Software engineers can develop AI for social good.

- Example: An engineer working in automotive design can focus on creating energy-efficient electric vehicles

(EVs) rather than just optimizing cars for speed and luxury.

Action Step: Find one challenge in your industry where you can contribute to a better future.

3. Build a Knowledge-Sharing & Mentorship Culture

True growth is not just about what you achieve, but also about how many people you uplift. The best engineers create a ripple effect by sharing their knowledge.

- Ways to mentor and give back:

  o Guide young engineers through mentorship programs.

  o Share expertise through blogs, books, or YouTube.

  o Teach real-world problem-solving through online courses.

- Example: An AI engineer can create a mentorship group for young professionals to learn about AI ethics and responsible AI development.

Action Step: Start sharing your knowledge—write a blog, record a video, or mentor someone today.

4.Create Sustainable and Ethical Engineering Solutions

Engineering decisions shape the world. With AI, automation, and new technologies rising, engineers must ensure ethical and sustainable solutions.

- Ethical engineering focus areas:

  - AI engineers must prevent bias in algorithms.

  - Civil engineers must prioritize eco-friendly materials.

  - Biomedical engineers must consider affordable healthcare solutions.

- Example: A software engineer working in AI can advocate for transparency in machine learning models instead of focusing only on profit-driven AI solutions.

Action Step: Identify one ethical challenge in your field and commit to a responsible approach.

5.Grow Spiritually & Emotionally Alongside Professionally

Growth is incomplete without inner fulfillment. Engineers should not just build external structures (products, machines, code) but also internal structures (values, mindset, purpose).

- Ways to nurture spiritual and emotional growth:

  - Practice mindfulness & reflection to stay balanced.

  - Engage in social impact projects for a sense of purpose.

  - Seek happiness beyond job titles and income.

- Example: A busy engineer who prioritizes mental well-being and work-life balance will be more creative, innovative, and fulfilled.

Action Step: Set one personal goal that nurtures your mind and soul, not just your career.

Engineers as Architects of a Better Future

To redefine growth beyond financial success, engineers must:

- Shift from success to significance

- Solve real-world problems that matter

- Share knowledge and mentor the next generation

- Create sustainable & ethical solutions

- Balance professional, emotional, and spiritual growth

b). Contributing to Society Through Innovation & Sustainable Engineering

As engineers, we have the power to create solutions that not only drive economic growth but also serve humanity and protect our planet.

True engineering success is not just about innovation but about responsible innovation—solutions that are sustainable, ethical, and impactful.

So, how can engineers actively contribute to society through innovation and sustainable engineering? Let's break it down.

1. Innovate with Purpose: Solve Real-World Challenges

Engineering is not just about developing new technologies—it's about creating solutions that address critical problems. Instead of following trends, engineers must design innovations that serve humanity.

- Key Areas for Purpose-Driven Innovation:

  - Sustainable Energy – Develop solar, wind, or bioenergy solutions.

  - Water Conservation – Improve wastewater management & purification.

  - Smart Infrastructure – Build resilient cities & eco-friendly materials.

  - Healthcare Technology – Create affordable medical devices.

  - AI for Good – Use AI to solve social & environmental challenges.

- Example: Instead of designing faster cars, an automotive engineer can focus on reducing emissions and increasing fuel efficiency.

Action Step: Identify one major problem in your field and think of an innovative, impactful solution.

2. Engineer for Sustainability: Reduce Environmental Impact

Sustainability is no longer an option—it's a necessity. Every engineer must integrate sustainability into their projects, whether designing a product, developing software, or building infrastructure.

- How Engineers Can Build Sustainably:

    o Use recyclable & biodegradable materials.

    o Design energy-efficient products & systems.

    o Follow green building standards for construction.

    o Optimize AI & computing to reduce energy waste.

    o Support the circular economy by designing for reuse.

- Example: A software engineer can optimize AI algorithms to use less computing power, reducing energy consumption in data centers.

Action Step: Look at your current projects and find one way to reduce waste, energy use, or environmental harm.

3. Make Engineering Accessible: Innovation for All

A truly impactful innovation should not just be for the privileged few—it should benefit everyone, including underprivileged communities.

- Ways to Make Engineering More Inclusive:

o Develop low-cost, high-impact solutions for rural areas.

o Improve connectivity & digital access for underserved communities.

o Build affordable healthcare technologies for mass use.

o Create energy-efficient solutions that lower costs for all.

- Example: A biomedical engineer designing prosthetic limbs can create low-cost 3D-printed versions for people in developing nations.

Action Step: Identify a gap in accessibility in your industry and think about how to make solutions more inclusive.

4. Promote Ethical & Responsible Engineering

With the rise of AI, automation, and biotechnology, engineers hold immense responsibility—not just to innovate but to ensure that innovation is ethical, fair, and safe.

- Principles of Ethical Engineering:

o Transparency – Ensure AI models & decisions are explainable.

o Fairness – Avoid bias in algorithms & data collection.

o Security – Protect user data & privacy.

- o Sustainability – Ensure long-term safety of tech solutions.

- • Example: An AI engineer developing facial recognition software must actively prevent biases that discriminate against certain ethnicities or demographics.

Action Step: Evaluate your work—does it align with ethical principles? If not, what changes can you make?

5. Collaborate & Inspire: Engineering as a Community Effort

True innovation doesn't happen in isolation. The best solutions come when engineers collaborate across disciplines and inspire the next generation.

- • Ways to Foster Innovation Through Collaboration:

  - o Work with scientists, policymakers & entrepreneurs.

  - o Share ideas through conferences, open-source projects & hackathons.

  - o Mentor young engineers & students to build a culture of innovation.

- • Example: A group of engineers working together on open-source climate change solutions can accelerate progress faster than working alone.

Action Step: Find one group, project, or community to contribute your expertise.

Engineers as Catalysts for a Better Tomorrow

To contribute meaningfully to society, engineers must:

- Innovate with purpose – Solve real-world challenges.

- Build for sustainability – Reduce environmental impact.

- Make technology accessible – Ensure solutions reach all.

- Adopt ethical practices – Innovate responsibly.

- Collaborate & mentor – Inspire the next wave of engineers.

Engineering is not just about progress—it's about progress with purpose.

Are you ready to redefine your role as an engineer for the betterment of society?

c) The Role of AI in Shaping the Future of Engineering Careers

AI is no longer just a tool— The Game Changer for Engineers,

it's transforming engineering. Whether you work in software, mechanical, civil, or any other engineering field, AI is redefining how engineers work, innovate, and grow.

If you don't embrace AI, you risk becoming obsolete. But if you learn to leverage AI, you can accelerate your career like never before.

Let's explore how AI is shaping the future of engineering and what you can do to stay ahead.

1. AI is Automating Repetitive Tasks, Freeing Engineers for Innovation

AI is replacing repetitive, time-consuming tasks, allowing engineers to focus on high-value work.

- How AI is Automating Engineering Tasks:

    o Data Analysis & Simulations – AI analyzes massive datasets faster than humans.

    o Structural Design & Optimization – AI optimizes building, machine, and circuit designs.

    o Coding & Debugging – AI-powered tools write, review, and debug code automatically.

    o Predictive Maintenance – AI detects equipment failures before they happen.

- Example: Civil engineers use AI to simulate earthquake resistance in buildings, reducing design errors and improving safety.

What You Can Do: Identify one task in your work that AI can automate and start using an AI tool to increase efficiency.

2. AI is Enhancing Design & Problem-Solving Through Generative AI

Generative AI is revolutionizing the way engineers design and innovate. Instead of manually creating solutions, AI generates multiple optimized designs based on given constraints.

- Key AI-Powered Design Tools:

    - Autodesk Generative Design – Creates optimized engineering models.

    - ANSYS AI Simulation – Assists in predictive engineering simulations.

    - DeepMind AlphaFold – Aids in biomedical engineering & protein folding solutions.

    - ChatGPT & CoPilot – Supports coding & documentation.

- Example: Automotive engineers use AI to create lighter, stronger, and more fuel-efficient vehicles by testing thousands of design variations.

What You Can Do: Learn to use an AI-powered design tool relevant to your field and start incorporating AI-generated solutions into your work.

3. AI is Creating New Engineering Roles & Specializations

AI is not taking away engineering jobs—it's creating new opportunities for those who adapt.

- Emerging AI-Driven Engineering Roles:

  o AI & Robotics Engineer – Designing intelligent automation systems.

  o AI-Powered Product Designer – Creating AI-integrated products.

  o Data & AI Engineer – Working on big data & machine learning models.

  o Sustainable Engineering with AI – Using AI for smart grids, water conservation, and climate solutions.

- Example: A mechanical engineer can transition into AI-driven predictive maintenance, reducing downtime in manufacturing industries.

What You Can Do: Identify an AI-driven niche in your field and start developing expertise in that area.

4. AI is Changing the Way Engineers Learn & Upskill

AI is making learning faster, smarter, and more personalized. Engineers no longer need to follow traditional learning paths—they can now use AI-driven platforms to learn exactly what they need, when they need it.

- AI-Powered Learning Platforms for Engineers:

  o Coursera & Udacity AI Courses – Personalized AI-driven course recommendations.

o   LinkedIn Learning with AI – Suggests career-specific learning paths.

o   ChatGPT & AI Tutors – Explain concepts, generate code, and assist in projects.

- Example: Engineers can use AI tutors to debug code, learn new programming languages, and solve complex equations in real time.

What You Can Do: Use AI-based learning platforms to upskill yourself in AI-driven engineering tools and techniques.

5. AI is Reshaping the Job Market: Engineers Must Adapt or Become Obsolete

The engineers of the future must learn to work alongside AI. If you resist AI, you risk becoming irrelevant. But if you adapt, you can accelerate your career growth.

- Future-Proof Your Engineering Career by Learning:

o   AI & Machine Learning – Even if you're not a software engineer, basic AI knowledge is crucial.

o   IoT & Smart Systems – AI is integrating with hardware & real-world systems.

o   AI-Driven Engineering Design – Learn AI-based CAD, simulation, and prototyping.

o   Data Analytics for Engineers – AI thrives on data—engineers who can analyze & interpret data will have an edge.

- Example: A civil engineer who understands AI-driven smart city design will be far ahead of those still using traditional methods.

What You Can Do: Identify one AI-driven skill that aligns with your engineering field and start learning it today.

AI is Not Replacing Engineers, It's Empowering Them

AI isn't here to take your job—it's here to make you more powerful. Engineers who embrace AI will lead the future, while those who ignore it may struggle to stay relevant.

- AI automates repetitive tasks, allowing engineers to focus on innovation.

- Generative AI enhances design efficiency and problem-solving.

- AI is creating new job opportunities—engineers must upskill to stay ahead.

- AI-driven learning platforms help engineers acquire the latest skills.

- The future belongs to engineers who embrace AI as a collaborator, not a competitor.

The question is: Will you be the engineer who resists AI or the one who thrives with it?

CONCLUSION

i. The Engineer's Roadmap for Continuous Growth

As an engineer, your growth doesn't stop after getting a degree or landing a job. True success comes from continuous learning, adapting, and evolving with the changing world. Whether you're in core engineering, software, AI, or any other field, this roadmap will help you stay ahead, master your expertise, and create a lasting impact.

Step 1: Master Your Expertise (Determine, Develop, Deploy)

1. Determine Your Expertise

Before you grow, define your niche and understand your strengths.

- Conduct a self-analysis: What are your strengths, interests, and market opportunities?

- Identify high-demand engineering skills that align with your passion.

- Validate your niche using the 10 Activity Framework & 5P Method.

- Create a strategy sheet to get clarity on your growth path.

Action Step: Write down your top three strengths and three areas where you need improvement.

2. Develop Your Expertise

Once you've identified your niche, it's time to learn, practice, and refine.

- Learn from mentors and industry leaders.

- Work on real-world projects to gain hands-on experience.

- Teach and share your expertise to reinforce your learning.

- Gather feedback to improve your skills and credibility.

Action Step: Start a mini-project in your niche and document your learnings.

3. Deploy Your Expertise

Turn your knowledge into real-world impact.

- Create products, courses, or consulting services based on your expertise.

- Learn packaging, pricing, and delivery strategies to scale your knowledge.

- Use AI and automation tools to streamline content creation & product delivery.

- Set up an LMS (Learning Management System) or online training program.

Action Step: Identify one monetizable skill and brainstorm how to package it as a product/service.

Step 2: Communicate Your Expertise (Saying, Selling, Serving)

4. Say – Build Your Personal Brand

Your knowledge is valuable only if people know about it.

- Create a strong online presence through social media, blogs, and a website.

- Use AI-powered content marketing & SEO to reach a wider audience.

- Engage in networking, public speaking, and webinars to establish authority.

Action Step: Write a LinkedIn post or a blog article about your expertise today.

5. Sell – Monetize Your Expertise

Turn your knowledge into income streams.

- Use value-based selling strategies to attract the right audience.

- Master persuasive communication & storytelling to sell with impact.

- Automate sales with AI-driven customer engagement tools.

Action Step: Create a simple offer (consulting, course, or service) and start marketing it.

6. Serve – Build Long-Term Relationships

Success isn't just about making sales—it's about delivering value consistently.

- Offer post-sale support & follow-ups to build trust.

- Use AI-powered chatbots & automation for customer service.

- Provide digital resources (PDFs, videos, and community access) to add value.

Action Step: Identify one way to improve your customer experience and implement it.

Step 3: Monetize & Scale (Make, Manage, Multiply Money)

7. Make Money – Get Paid for Your Expertise

Engineers are natural problem solvers—and solving problems creates income.

- Identify high-value problems in your industry.

- Create solutions that people are willing to pay for.

- Set up multiple monetization methods: products, courses, consulting, or services.

Action Step: Brainstorm one income-generating idea based on your expertise.

8. Manage Money – Build Financial Discipline

Managing money is as important as making it.

- Categorize your income into different financial accounts for stability.

- Use AI-based financial tools to track earnings and expenses.

- Develop money management habits to sustain long-term success.

Action Step: Set up separate bank accounts for business, savings, and investments.

9. Multiply Money – Scale Your Impact

To grow exponentially, diversify your income and scale your expertise.

- Explore multiple income streams (affiliate marketing, e-commerce, partnerships).

- Scale your job or business using AI and automation.

- Expand into cross-selling, up-selling, and product bundling strategies.

Action Step: List down three ways you can scale your current income streams.

Engineering a Future of Impact & Innovation

10. Engineers as Architects of the Future

Growth is not just about money—it's about creating a lasting impact.

- Use your skills to contribute to innovation & sustainability.

- Build solutions that help society and future generations.

- Stay adaptable—continuous learning is the key to long-term success.

Action Step: Write down how you want to contribute to the world as an engineer.

The Engineer's Continuous Growth Roadmap

Expertize : Determine, Develop, and Deploy

Determine = Clarity → This phase is about gaining clarity on your expertise. You assess your strengths, interests, and industry needs to identify the right niche where you can create impact.

Develop = Mastery → Once you have clarity, you build mastery by learning, applying, and refining your expertise. This involves mentorship, hands-on projects, feedback, and continuous improvement.

Deploy = Placement → After mastering your expertise, you focus on positioning your solutions at the right time, in the right place, and in the right form.

This means designing the right product, selecting the best delivery method, and ensuring it reaches the right audience effectively.

DETERMINE → CLARITY (Find your niche & purpose)

DEVELOP → MASTERY (Learn, apply, and refine skills)

DEPLOY → PLACEMENT (Strategically position your expertise for maximum impact)

By following this approach, engineers can ensure their expertise is not just developed but also placed effectively in the market, setting the stage for monetization and long-term success.

Communicate: Say, Sell, and Serve

Say = Value → You communicate your expertise by sharing valuable insights, educating, and building trust.

Sell = Transformation → You are not just selling a product or service; you are selling a solution that transforms your customer's situation. The focus is on solving problems and creating impact.

Serve = Growth → After selling, you ensure customer success by providing support, guidance, and continuous value. Serving builds long-term relationships, trust, and referrals, leading to sustainable growth for both you and your clients.

SAY → VALUE (Educate & Build Authority)

SELL → TRANSFORMATION (Solve & Deliver Results)

SERVE → GROWTH (Support & Retain for Long-Term Success)

This ensures engineers not only establish their expertise ,but communicate to monetize and scale it effectively

Monetize: Make, Manage, and Multiply

Make = Earning → This phase is about more than just creating value—it's about effectively communicating that value and earning money.

Engineers must learn how to position their expertise, attract the right audience, and prompt them to take action and pay for solutions.

Manage = Financial Discipline → Once you start earning, structured money management becomes essential.

Categorizing income, optimizing spending, and setting up financial systems help maintain stability and long-term wealth.

Multiply = Wealth Expansion → After managing finances effectively, the next step is to scale and multiply income through smart investments, business expansion, cross-selling, up-selling, automation, and passive income streams.

MAKE → EARN (Communicate value, get paid)

MANAGE → CONTROL (Structure finances, ensure stability)

MULTIPLY → SCALE (Expand income, grow wealth)

By following this structured approach, engineers can turn their expertise into sustainable and scalable financial success

Create a Meaningful Legacy

Engineers as Architects of the Future: Innovate, contribute, and redefine success.

Engineering success is not a one-time achievement—it's a journey of continuous learning, adaptation, and impact.

ii. Next Steps to Implement Expertise-Building Strategies

Now that you have a structured roadmap for continuous growth, it's time to take action.

Follow these step-by-step implementation strategies to build your expertise effectively and monetize your knowledge.

Step 1: Self-Assessment & Clarity (1-Week Plan)

Goal: Identify your strengths, niche, and growth path.

Action Steps:

- Conduct a self-audit: List your technical & soft skills.
- Use the 10 Activity Framework , 5P &5Why Method to validate your microniche.
- Create a strategy sheet outlining your expertise focus.
- Define your long-term purpose & vision for impact.

Task: Write a one-page clarity document on your expertise & goals.

Step 2: Structured Learning & Application (4-Week Plan)

Goal: Master your skills through structured learning & real-world application.

Action Steps:

- Find three top mentors in your industry (books, courses, or direct mentors).

- Enroll in structured online courses (Coursera, Udemy, LinkedIn Learning, AI-based learning tools).

- Start one real-world project related to your niche.

- Teach what you learn through articles, videos, or social media posts.

Task: Publish a small tutorial or case study on what you've learned.

Step 3: Creating & Monetizing Knowledge (6-Week Plan)

Goal: Convert your expertise into products, services, or consulting opportunities.

Action Steps:

- Choose a digital product (e.g., ebook, online course, consulting package).

- Design your first offer (Pricing, Packaging, Promotion).

- Use AI for content creation (ChatGPT, Jasper, Canva, AI video tools).

- Set up a simple website or landing page to showcase your expertise.

Task: Launch a beta version of your digital product/service and test with 5-10 users.

Step 4: Building a Personal Brand & Audience (8-Week Plan)

Goal: Establish yourself as an authority in your niche & attract the right audience.

Action Steps:

- Optimize your LinkedIn profile & website for credibility.

- Start posting regularly on social media (LinkedIn, Twitter, YouTube).

- Leverage AI-based content marketing & SEO to reach more people.

- Engage in networking, webinars, and industry events.

Task: Grow your email list & social media audience to at least 500 engaged followers.

Step 5: Scaling & Automation (Ongoing Strategy)

Goal: Expand your impact, generate recurring income, and automate processes.

Action Steps:

- Explore AI-driven sales automation & chatbots for customer engagement.

- Implement passive income streams (affiliate marketing, subscriptions, memberships).

- Use AI-powered financial management tools to track income & expenses.

- Continuously update and refine your offerings based on customer feedback.

Task: Set up at least one automated income stream within the next three months.

Take Action & Track Progress!

Your expertise-building journey requires consistent action, learning, and iteration. Start today, track your progress, and refine along the way!

iii. Joining a Community of Expert Engineers

Success in engineering isn't just about what you know—it's also about who you connect with. A strong community of like-minded engineers can accelerate your growth, provide valuable insights, and open doors to new opportunities.

Here's how you can find, join, and actively participate in expert engineering communities to fuel your career growth.

Why Join an Expert Engineering Community?

- Networking – Connect with top industry professionals, mentors, and peers.

- Knowledge Sharing – Stay updated with the latest trends, AI-driven technologies, and industry innovations.

- Opportunities – Access job openings, collaborations, and potential clients.

- Support & Motivation – Gain guidance, feedback, and encouragement from like-minded individuals.

- Growth & Monetization – Learn how to leverage your expertise for financial success.

Types of Engineering Communities to Join

1. Online Communities & Forums

- LinkedIn Groups – Engineering Leadership, AI in Engineering, Freelance Engineers

- Reddit – r/engineering, r/AskEngineers, r/MachineLearning

- Facebook Groups – AI & Engineering, Mechanical Engineers Network, Engineering Entrepreneurs

- Quora Spaces – Engineering Excellence, AI & Automation

2. Professional Associations & Societies

- IEEE (Institute of Electrical and Electronics Engineers)
- ASME (American Society of Mechanical Engineers)
- ACI (American Concrete Institute) – Civil Engineering
- SAE International – Automotive Engineers
- The Institution of Engineers (India) (IEI)
- Indian Society for Technical Education (ISTE)
- Engineering Council of India (ECI)
- Indian Institution of Industrial Engineering (IIIE)
- Indian Concrete Institute (ICI)
- Indian Roads Congress (IRC)
- Institution of Electronics and Telecommunication Engineers (IETE)
- IEEE India Section
- Computer Society of India (CSI)
- Association for Computing Machinery (ACM) India
- Indian Institute of Chemical Engineers (IIChE)
- Aeronautical Society of India (AeSI)
- Society of Automotive Engineers (SAE) India

3. Industry-Specific Events & Meetups

- Webinars & Conferences – Attend AI, automation, and engineering summits.

- Meetup.com – Find local engineering, AI, and startup events.

- Hackathons & Competitions – Participate in AI, robotics, or automation challenges.

4. Private Mastermind Groups & Coaching Programs

- Join exclusive mastermind groups for engineers focusing on career growth.

- Enroll in mentorship programs where industry leaders guide you.

- Be part of a coaching program that provides step-by-step growth strategies.

How to Actively Participate & Get the Most Value

- Engage Regularly – Ask questions, share insights, and contribute to discussions.

- Share Your Work – Post case studies, project experiences, or technical insights.

- Offer Value First – Help others by solving problems, offering solutions, or mentoring.

- Build Meaningful Connections – Network with professionals, collaborate, and find mentors.

- Stay Consistent – Make community engagement a habit for continuous learning & growth

The world is changing at an unprecedented pace, and as an engineer, you have two choices—adapt and thrive or resist and struggle. The knowledge, strategies, and frameworks in this book are not just theories; they are a roadmap to transform your career, finances, and life.

But knowledge alone isn't enough. The real transformation happens when you take action. Implement what you've learned, refine your unique expertise, communicate powerfully, and monetize effectively. Engineering is no longer just about solving problems; it's about creating value, influencing change, and securing your financial future.

Now, it's time to step up. Surround yourself with those who challenge, inspire, and uplift you. Your growth journey doesn't end here—it begins here.

I wish you a life of abundance, joy, and freedom.

Now go out there and make it happen!

With all my heart,

Pradeepkumar K Padmanabhan

Book Summary

Introduction:

The engineering profession is undergoing a significant transformation. Traditional career paths are becoming less reliable, and the rise of AI and automation is reshaping the job market.

Engineers who want to thrive in this new era must redefine growth—not just in terms of technical skills but in how they position themselves as experts, communicate their value, and create sustainable income streams.

This book provides a structured approach to navigating this shift. It introduces a framework that helps engineers build specialized expertise, leverage communication for career advancement, and implement monetization strategies that go beyond conventional employment. By following this roadmap, engineers can achieve financial independence while making a meaningful impact.

EXPERTIZE:

Expertise is the foundation of career success in an AI-driven world. Engineers must move beyond general knowledge and develop specialized skills that solve high-value problems. This section of the book breaks expertise development into three key phases: determining a niche, developing in-depth knowledge, and deploying it effectively.

The process begins with self-analysis to identify strengths, interests, and market opportunities. Engineers must define a clear career vision and align their expertise with industry needs.

Using structured frameworks like the 10 Activity Framework, the 5P Method, and the 5WHY Analysis, readers can systematically choose a high-impact niche.

Once a niche is defined, the next step is skill development. This involves structured learning, mentorship, and hands-on application through projects. Teaching and sharing knowledge further reinforce expertise and build credibility. AI-powered learning platforms can accelerate this process, providing personalized learning paths and real-time feedback.

The final step is deploying expertise. Engineers must package their knowledge into products, services, or content that demonstrate their value.

Strategies for creating educational materials, consulting services, and AI-driven solutions are explored, along with tools for automating content creation and distribution. This ensures that expertise is not just developed but also recognized and monetized.

COMMUNICATE:

Technical skills alone are no longer enough to stand out. Engineers must effectively communicate their expertise to build authority and attract opportunities. This section emphasizes the importance of strategic visibility and engagement.

The first aspect of communication is establishing an online presence. Social media, personal websites, and blogs play a critical role in positioning engineers as thought leaders. AI-powered content marketing and SEO strategies help amplify reach and engagement. By consistently sharing insights, engineers can attract the right audience and build professional credibility.

Beyond visibility, effective communication is essential for converting knowledge into business opportunities. Engineers must understand customer psychology, trust-building techniques, and value-based sales strategies. Persuasive storytelling and clear messaging enhance the impact of their communication. AI-driven tools further optimize sales and customer engagement, enabling efficient and scalable outreach.

Communication is not just about promotion—it is also about long-term relationship building. Delivering consistent value through high-quality customer service, AI-powered automation, and digital resources strengthens client trust and retention.

Engineers who excel in this area create sustainable, long-term professional growth.

MONETIZE:

Many engineers rely solely on salaries without realizing the potential of their expertise as a revenue source. This section provides a structured approach to monetizing technical knowledge and transitioning from job-based income to expertise-based income.

The first step in monetization is identifying high-value problems to solve. Engineers must analyze industry needs and create solutions that provide measurable value. Business models such as consulting, digital products, online courses, and AI-driven services offer multiple pathways for income generation. The book provides practical strategies for packaging, pricing, and delivering these solutions effectively.

Beyond making money, managing income strategically is essential for financial stability. Engineers are introduced to AI-driven financial management tools, income categorization techniques, and habits that contribute to long-term wealth-building. The focus is on creating financial systems that ensure consistent growth and security.

The final aspect of monetization is scaling. Engineers can expand their income through multiple streams, including cross-selling, up-selling, affiliate marketing, and AI-driven automation. Digital products, e-commerce solutions, and scalable service models enable continuous revenue growth. By leveraging technology, engineers can maximize their financial potential while maintaining work-life balance.

BEYOND YOUR SUCCESS:

Success is not just about financial growth—it is also about impact. Engineers have the potential to contribute to society through innovation and sustainable solutions. This book encourages professionals to think beyond individual achievements and explore ways to create lasting change through their expertise.

By integrating AI, technology, and ethical business practices, engineers can lead advancements that benefit industries and communities. The final section of the book outlines a roadmap for continuous learning, adaptation, and growth. It also emphasizes the importance of networking and collaboration, encouraging engineers to join professional communities that support knowledge exchange and collective progress.

Conclusion:

The engineering profession is evolving, and those who adapt will thrive. This book provides a step-by-step guide to developing expertise, mastering communication, and implementing monetization strategies that create long-term success. By following these principles, engineers can achieve financial independence, career fulfillment, and a meaningful impact in their field.

The journey to growth is continuous, and this book serves as a practical companion for engineers committed to lifelong learning and success.

Acknowledgment

With deep gratitude and reverence, I express my heartfelt appreciation to all those who have been a guiding light on my journey, leading to the creation of this book.

I bow in gratitude to my beloved parents, M. Padmanabhan Nair and K. Devaky Nessiar—even though they are not physically present, their blessings, values, and unconditional love continue to guide and inspire me every moment. Their presence is felt in every step I take.

My heartfelt gratitude to all my engineering teachers, seniors, mentors, classmates, and students who have taught me engineering—both formally and informally. Your guidance, support, and shared knowledge have been invaluable in shaping my journey.

I hold immense admiration for my role model, Dr. APJ Abdul Kalam, whose life and vision continue to inspire me.

I deeply respect my spiritual Gurus and sincerely thank my New Age trainers and coaches, especially Siddharth Rajsekar, for his guidance in digital coaching.

To my wife, Cinderella, my unwavering pillar of support, and to my children, Adhree and Niyatee, for their love, patience, and encouragement—I am forever grateful. My brother, sister, uncles, family members, and dear friends have stood by me, making this journey meaningful.

I also acknowledge Som Bathla and Ravi Lalit Tewari, who have been invaluable coaches and guides in my journey as an author.

Every person I have met, every challenge I have faced, and every book I have read has contributed to my learning and growth. These experiences, combined with the wisdom of countless mentors, have led to the birth of this book.

I must acknowledge the pivotal role of Technology and AI tools like ChatGPT in shaping this book and accelerating its publication. These innovations have been invaluable thinking partners, refining my ideas, enhancing my expression, and making the journey of bringing this book to life both efficient and enriching.

Above all, I express my deepest gratitude to the universal power that signalled my intellect to embark on this journey. Without this predesigned guidance, this book would not have been possible.

I extend my sincere thanks to the publishers of this book for bringing my work to the world and to my students, whose curiosity, dedication, and pursuit of growth continue to inspire me every day.

This book is a humble offering to all engineers' growth and success. May it serve as a guiding light in your journey to redefine your career growth. Jai Hind!

ABOUT THE AUTHOR

Pradeepkumar K Padmanabhan | Former Indian Army Officer | Engineer | Techpreneur | Author | Engineers' Growth Coach|

Pradeepkumar K Padmanabhan (Capt Pradeepkumar KP, Retd.) is a former Indian Army Officer, seasoned Engineer, Techpreneur, and Engineers' Growth Coach with over 30 years of diverse industry experience. From serving as a Technical Officer in the Indian Army to spearheading multiple entrepreneurial ventures,

Now, his true calling is empowering engineers—helping them build purpose-driven careers, achieve financial success, and create a balanced life.

The TRUE Mission – Transforming Engineers to Their Best

Throughout his career, Pradeepkumar observed that many engineers struggle with job dissatisfaction, low performance, and unemployment—primarily due to mindset misalignment. He realized that success comes from aligning engineers with their profession and unlocking the potential of underutilized technical resources.

This insight led him to establish the TRUE Growth Model, creating a new breed of professionals known as Technical Resource Utilization Experts (TRUE)®. His mission is to transform engineers into authentic, competitive, and responsible leaders—whether in a job or business.

Through his proven strategies, he helps engineers:

Expertize – Master technical, communication, and leadership skills

Communicate – Build influence, confidence, and credibility

Monetize – Leverage skills and knowledge to create high-income opportunities

Innovate with Integrity – Drive sustainable success while making a meaningful impact

Join the Movement – Engineer Your Success!

Pradeepkumar K Padmanabhan is on a mission to help 1,000,000 engineers break free from limitations and achieve high growth, professional fulfillment, and happy living.

His book, Engineers' Growth Redefined, reflects this concepts —offering a structured, systematic and result oriented approach for engineers to expertize, communicate, monetize their skills and knowledge in AI era.

Through his coaching and mentorship, he empowers engineers to become TRUE; Technical Resource Utilization Experts, and to achieve prosperity, freedom and purpose with a successful life.

To know more details.

Email: pradeepdesign@gmail.com

Website: www.pradeepkumarkp.com

Redefine your growth as an engineer!

DISCLAIMER

The information provided in this book is for general informational purposes only.

While every effort has been made to ensure the accuracy and reliability of the content at the time of publication, the author and publisher make no representations or warranties of any kind, express or implied, about the completeness, accuracy, or suitability of the information contained herein for any purpose.

This book is not intended to serve as a substitute for professional advice, including but not limited to legal, medical, financial, or other specialized guidance.

Readers are encouraged to consult with qualified professionals regarding their specific circumstances before making decisions based on the information in this book.

The opinions expressed in this book are those of the author and do not necessarily reflect the views of the publisher or any other affiliated parties.

Any references to real events, people, or entities are included for illustrative purposes and should not be interpreted as endorsements or definitive statements of fact.

The author and publisher shall not be liable for any loss, damage, or inconvenience arising from the use of, or reliance on, the information contained in this book. By reading this book, you acknowledge that you assume full responsibility for your actions and decisions based on its content.

May I Ask You For A Small Favor?

First, I want to thank you for reading this book. You could have chosen any other book, but you took mine, and I appreciate this. I hope you have at least a few actionable insights that will positively impact your daily life.

Can I ask for 30 seconds more of your time?

I'd love it if you could leave a review of the book. That will help me grow my readership by encouraging folks to take a chance on my books.

Keeping it straight - reviews are the lifeblood of any author.

It will take less than a minute of your time but will tremendously help me reach out to more people.

If you liked this book, please consider posting an honest review on your preferred retailer. And I'd love to see your review. Thanks for your support.